GIS Applications in Green Development

GIS Applications in Green Development

Editor

Yannis Maniatis

MDPI

Basel • Beijing • Wuhan • Barcelona • Belgrade • Novi Sad • Cluj • Manchester

Editor
Yannis Maniatis
Department of
Digital Systems
University of Piraeus
Piraeus
Greece

Editorial Office
MDPI
St. Alban-Anlage 66
4052 Basel, Switzerland

This is a reprint of articles from the Special Issue published online in the open access journal *Applied Sciences* (ISSN 2076-3417) (available at: https://www.mdpi.com/journal/applsci/special_issues/GIS_Green_Development).

For citation purposes, cite each article independently as indicated on the article page online and as indicated below:

Lastname, A.A.; Lastname, B.B. Article Title. *Journal Name* **Year**, *Volume Number*, Page Range.

ISBN 978-3-7258-0649-2 (Hbk)
ISBN 978-3-7258-0650-8 (PDF)
doi.org/10.3390/books978-3-7258-0650-8

Contents

applied sciences

Editorial

Special Issue "GIS Applications in Green Development"

Yannis Maniatis

Department of Digital Systems, University of Piraeus, 18534 Piraeus, Greece; maniatis@unipi.gr

In the context of climate change, the role of Geographic Information Systems (GIS) in green developments cannot be overstated. The application of smart GIS is the linchpin for decision makers tasked with designing and monitoring climate-conscious solutions at local, national, and international scales. With 75% of the Earth's surface now impacted by human activities, it is imperative to expand the use of smart GIS to predict and mitigate the impact of these activities across forests, oceans, urban and rural areas, transportation networks, and production sites. This Special Issue of Applied Sciences, titled "GIS Applications in Green Development", explores the pivotal role of GIS in advancing sustainability across diverse domains.

The five papers presented in this Special Issue consider the potential intersection of GIS and green developments in urban planning, renewable energy integration, disaster management, and the energy sector. According to Ostapenko et al. [1], the potential to implement renewable energy sources in Ukraine is scrutinized using global and local Geographic Information Systems (GIS). The study highlights GIS's prowess in identifying suitable territories for renewable energy development, assessing technical potential and facilitating the integration of renewable energy technologies in Ukraine's energy sector. Zorzano-Alba et al. [2] addressed the sensitive issue of the visual impact associated with renewable energy infrastructure, introducing a novel methodology for identifying optimal locations for photovoltaic power plants, especially in areas of cultural or scenic significance. Maniatis et al. [3] focused on fire risk mapping in the context of climate change. The authors presented an innovative approach, incorporating recent land cover changes, to highlight regions with a high fire risk. Through the integration of a support vector machine (SVM) algorithm and the analytic hierarchy process (AHP) within a GIS framework, the authors created a robust fire risk estimation model. The model identifies high-risk areas in the Dadia-Lefkimi-Soufli National Forest Park, Greece, (although it can be adapted for other regions) reinforcing the vital role of GIS in disaster management. Pinna et al. [4] offer a comprehensive assessment of Sardinia's rooftop photovoltaic potential using GIS data and an efficient shadow calculation algorithm. Their innovative approach provides a high-resolution, full census evaluation of the photovoltaic potential, which can be applied on a regional scale. By estimating not only the geographic but also the technical and economic potential, the paper exemplifies how GIS facilitate large-scale renewable energy planning. Yildiz [5] explores the wind energy potential of Balıkesir Province, Turkey, through GIS functions. The study employs wind speed data from meteorological stations and extrapolates it to create a wind speed map, enhancing this methodology by using an equation for turbine placement that is compliant with national regulations. This innovative approach enables the calculation of wind energy potential across the province, contributing to the knowledge regarding renewable energy assessments using GIS.

The collection of papers in this Special Issue emphasizes that GIS are more than a technology; in fact, they are an indispensable tool in the quest for green developments and sustainable management. By providing insights, data-driven decision support, and innovative methodologies, GIS empower us to address the profound environmental challenges of our time.

Citation: Maniatis, Y. Special Issue "GIS Applications in Green Development". *Appl. Sci.* **2023**, *13*, 10856. https://doi.org/10.3390/app131910856

Received: 25 September 2023
Accepted: 27 September 2023
Published: 29 September 2023

Acknowledgments: I would like to express my gratitude to the authors, reviewers, and the Applied Sciences team for their contributions to this Special Issue.

Conflicts of Interest: The author declares no conflict of interest.

References

1. Ostapenko, O.; Olczak, P.; Koval, V.; Hren, L.; Matuszewska, D.; Postupna, O. Application of Geoinformation Systems for Assessment of Effective Integration of Renewable Energy Technologies in the Energy Sector of Ukraine. *Appl. Sci.* **2022**, *12*, 592. [CrossRef]
2. Zorzano-Alba, E.; Fernandez-Jimenez, L.A.; Garcia-Garrido, E.; Lara-Santillan, P.M.; Falces, A.; Zorzano-Santamaria, P.J.; Capellan-Villacian, C.; Mendoza-Villena, M. Visibility Assessment of New Photovoltaic Power Plants in Areas with Special Landscape Value. *Appl. Sci.* **2022**, *12*, 703. [CrossRef]
3. Maniatis, Y.; Doganis, A.; Chatzigeorgiadis, M. Fire Risk Probability Mapping Using Machine Learning Tools and Multi-Criteria Decision Analysis in the GIS Environment: A Case Study in the National Park Forest Dadia-Lefkimi-Soufli, Greece. *Appl. Sci.* **2022**, *12*, 2938. [CrossRef]
4. Pinna, A.; Massidda, L. A Complete and High-Resolution Estimate of Sardinia's Rooftop Photovoltaic Potential. *Appl. Sci.* **2023**, *13*, 7. [CrossRef]
5. Yildiz, S.S. Determining Wind Energy Potential Using Geographic Information System Functions: A Case Study in Balıkesir, Turkey. *Appl. Sci.* **2023**, *13*, 9183. [CrossRef]

applied sciences

MDPI

Article

Determining Wind Energy Potential Using Geographic Information System Functions: A Case Study in Balıkesir, Turkey

Selim Serhan Yildiz

Department of Geomatics Engineering, Osmaniye Korkut Ata University, Osmaniye 80000, Turkey; serhan@osmaniye.edu.tr

Abstract: With developing technology, energy consumption and requirements are steadily rising. Wind energy emerges as an indispensable energy source in the world, where energy requirements are increasing gradually due to important features such as being renewable, sustainable, easily accessible, and environmentally friendly. In recent years, wind energy investments in Turkey have increased significantly, in line with the rest of the world. In recent decades, a significant number of investors have performed investment studies in this area. In this study, the wind energy potential of the Balıkesir Province in Turkey was calculated using geographical information system (GIS) functions. A wind speed map was created by using wind speed data measured at a 10 m altitude at 32 meteorological stations. The wind speeds were extrapolated to 100 m considering the land cover data, and a wind speed map of the Balıkesir Province was created using GIS functions. An equation was produced to calculate the number of turbines that can be placed in a certain area depending on the national regulation, which is also the novelty of this study. By using this equation, the wind energy potential values of the Balıkesir Province and its districts were obtained according to varying wind speed ranges. The results obtained in this study were compared with the wind energy potential atlas of Turkey (REPA).

Keywords: wind energy; wind speed; potential assessment; geographical information system (GIS); renewable energy

Citation: Yildiz, S.S. Determining Wind Energy Potential Using Geographic Information System Functions: A Case Study in Balıkesir, Turkey. *Appl. Sci.* **2023**, *13*, 9183. https://doi.org/10.3390/app13169183

Academic Editor: Yannis Maniatis

Received: 21 June 2023
Revised: 31 July 2023
Accepted: 3 August 2023
Published: 12 August 2023

1. Introduction

Energy consumption is increasing due to the rapidly growing world population and technological developments. Natural energy resources play an important role both in meeting the needs of countries and in development efforts throughout history. Today, countries that are advanced in terms of their development are those that have managed to use their natural resources efficiently. People want to benefit from all resources on Earth and have their needs met, which is difficult due to the rapid increase in global population. Planning and project studies about determining, obtaining, storing, processing, and using natural resources to meet the needs of people have shown how important natural resources are for human life. Natural energy resources include renewable and non-renewable sources. The main sources of non-renewable energy are oil, natural gas, coal, and nuclear energy. Oil, natural gas, and coal are called fossil fuels. Biomass, geothermal, solar, wind, and hydroelectric energy sources are the leading sources of renewable energy. Due to the lack of continuity of non-renewable energy resources, it is not possible to meet the energy needs of the world forever. According to Marks-Bielska et al. [1], experts warn that non-renewable energy sources may run out by 2040 with the present energy consumption in the world. The air pollution caused by the greenhouse gases generated during the consumption of fossil fuels poses a threat to the climate balance of the Earth [2]. Renewable energy is a type of energy source that depends on the climatic conditions of a region, produces less greenhouse gas emissions than fossil energy sources, and is naturally replenished [3].

Renewable energy sources are more advantageous than fossil energy sources due to their unlimited amount, less harm to the environment, and safety [4].

Solar and wind energy are the best known and fastest-growing renewable energy sources that will help to ensure sustainable development in the world [5]. Wind energy is the most developed and commercially available energy type among the renewable energy sources in the world. The People's Republic of China ranks first for the development of wind energy production in recent years. The United States, Germany, India, and Spain follow the People's Republic of China [6–9]. The installed wind power capacities of countries and their places in the world rankings are given in Table 1 [8].

Table 1. Installed capacity and ranks of countries [8].

Rank	Country	Installed Capacity 2020 (MW)	New Installations 2021 (MW)	Growth Rates (%)	Installed Capacity 2021 (MW)	Share in the World (%)
1	P.R. China	288,029	58,641	20.4	346,670	41.0
2	United States	122,328	12,518	10.2	134,846	15.9
3	Germany	62,208	1716	2.8	63,924	7.6
4	India	38,625	1732	4.5	40,357	4.8
5	Spain	27,446	750	2.7	28,196	3.3
6	United Kingdom	24,167	2645	10.9	26,812	3.2
7	Brazil	18,010	3827	21.2	21,837	2.6
8	France	17,949	1132	6.3	19,081	2.3
9	Canada	13,627	677	5.0	14,304	1.7
10	Sweden	9922	2175	21.9	12,097	1.4
11	Italy	10,850	258	2.4	11,108	1.3
12	Turkey	9305	1797	19.3	11,102	1.3
13	Others	99,992	12,964	13.0	112,956	13.4

Considering the Republic of Turkey's eleventh development plan covering the 2019–2023 period, the share of renewable energy sources in electricity production for 2023 is estimated to be 38.8% [10]. According to Turkey Electricity Investments in the June 2022 Summary Report prepared by the General Directorate of Energy Affairs, 54.05% of the total installed capacity of 101,518 MW, as of 30 June 2022, consists of renewable energy production facilities [11].

Wind energy production in Turkey first began in 1998 in Izmir [12]. After the initial setup, the development of wind energy production in Turkey has progressed in a way consistent with its general development in the world. However, the production of wind energy investments in Turkey has significantly increased in recent years. The total installed capacity, which was 1375.80 MW in 2010, reached 11,101.82 MW as of January 2022. The highest growth in terms of the installed capacity during this period was 1797 MW between 2020 and 2021. This increase elevated Turkey's place in the world rankings significantly. The annual rate of increase, which was around 30% in the first years, decreased noticeably between 2016 and 2019, and declined by around 9% in 2019. After 2019, the annual rate of increase started to rise again and reached 19.31% at the end of 2021. In recent years, in order to increase the renewable energy production in Turkey, a number of incentives, such as tax exemptions, fixed price guarantees, and public land allocation, have been given by the government [13]. With the help of these incentives, wind energy investments, which are an important renewable energy source, have gained momentum. When cities are compared in terms of their total installed capacity, İzmir ranks first with 1886.70 MW, Balıkesir second with 1375.05 MW, Çanakkale third with 917.35 MW, and Manisa fourth with 727.55 MW. The total installed capacity of these four cities constitute 44.3% of Turkey's installed capacity [14]. When the installed capacity increase of the provinces between January 2021 and January 2022 is examined, İstanbul takes the lead with an increase of 350.09 MW. Yalova, Çanakkale, Balıkesir, Kırklareli, Bursa, and Sakarya follow İstanbul [14,15]. The installed capacities, growth rates, and shares of the provinces in the country's capacity as of January 2021 and January 2022 are shown in Table 2. In addition, the geographical distribution of the provinces within the top twenty in terms of their installed wind power capacity in Turkey is given in Figure 1.

Table 2. The installed capacities, growth rates, and shares of the provinces in Turkey [14,15].

Rank	Province	Installed Capacity January 2022 (MW)	New Installations 2021 (MW)	Growth Rates (%)	Installed Capacity January 2021 (MW)	Share in the Country (%)
1	İzmir	1886.70	88.55	4.9	1798.15	17.0
2	Balıkesir	1375.05	155.00	12.7	1220.05	12.4
3	Çanakkale	917.35	166.03	22.1	751.32	8.3
4	Manisa	727.55	10.80	1.5	716.75	6.6
5	İstanbul	684.39	350.09	104.7	334.30	6.2
6	Kırklareli	481.68	150.08	45.3	331.60	4.3
7	Hatay	437.85	23.20	5.6	414.65	3.9
8	Afyon	368.45	0.00	0.0	368.45	3.3
9	Aydın	365.60	72.00	24.5	293.60	3.3
10	Konya	337.80	24.00	7.6	313.80	3.0
11	Yalova	308.95	174.60	130.0	134.35	2.8
12	Bursa	286.00	143.40	100.6	142.60	2.6
13	Kayseri	274.35	0.00	0.0	274.35	2.5
14	Osmaniye	265.30	0.00	0.0	265.30	2.4
15	Mersin	253.55	0.00	0.0	253.55	2.3
16	Muğla	237.25	32.00	15.6	205.25	2.1
17	Tekirdağ	187.95	10.25	5.8	177.70	1.7
18	Kırşehir	168.00	0.00	0.0	168.00	1.5
19	Sakarya	162.70	135.10	489.5	27.60	1.5
20	Sivas	155.30	0.00	0.0	155.30	1.4

Figure 1. Geographical distribution of the provinces within the top twenty in terms of installed wind power capacity in Turkey.

Wind speed and wind power potential are among the most important factors taken into consideration when establishing wind power plants. Therefore, the availability of standardized wind speed data is crucial for potential determination studies. Having up-to-date and reliable data in line with standards will enable consistent modeling procedures and comparatively reduce the effort spent on modeling [16].

The lowest (cut-in) speed required for current wind turbines to start generating electricity is generally 3 m/s [17–19]. Although wind turbines can produce energy at low speeds, wind speeds must be above a certain value for economically sustainable production. There are different studies about this subject in the literature. The exclusion value of wind speed considered in these studies varies between 5 and 7 m/s [20–27].

When the literature about determining wind energy potential is examined, studies are generally carried out based on only one point [28–38]. It is not very meaningful to make evaluations for large areas such as districts and provinces by using only one point. For example, Bertrand et al. [34] determined the energy density of Ambam in Cameroon. Wang and Liu [35] studied the wind energy potential of the Maling Mountain in China at a height of 30 m. The wind speeds were not extrapolated to a certain height. Ongaki et al. [39] determined the wind power density of the Kisii Region in Kenya using wind speed data obtained from four stations at a height of 10 m. The wind speeds were extrapolated to different heights from 20 m to 70 m. The wind power density of the case study's station was calculated in accordance with these wind speeds. Çakmakçı and Hüner [36] calculated the wind energy potential of the Kırklareli University Kayalı Campus at an altitude of 100 m. Paraschiv et al. [37] obtained the wind power density in the southeast of Romania near to Tulcea city. All used wind speeds obtained from only one station. Jung and Schindler [38] assessed the wind energy potential of Germany using several data points under climate change at the wind turbine scale. They produced a wind speed map by using a Wind Speed-Wind Shear model (WSWS). Although the wind speed map was created in the study, the wind potential evaluations were made on a point basis. Since the studies were carried out using a point-based approach, it is not meaningful to use this for the power potential calculation of large areas. Wind speed measured at multiple points spread over the studied area is required to consistently calculate the wind power potential over large areas.

With the help of geographic information system (GIS) functions, wind speed values at multiple points can be used as input data, and wind speed maps of a particular area can be created using interpolation methods. In this way, a basis is provided for studies in order to calculate the wind power potential in a specific area. Anwer and Deshmukh [40] created a wind speed map of the southern states of India based on 28 data points using the kriging interpolation method of GIS tools. The wind speed map was created in counters to illustrate the wind speed values. Since the produced map was black and white, it was not clearly understood which wind speed value the areas between the two counters had. Feng et al. [41] produced a wind speed map of mainland China using the inverse distance weighting (IDW) interpolation method. The wind energy potential of the study area was calculated for suitable areas for wind farm installation by excluding restricted areas, similar to this study. Zahedi et al. [42] adopted a multi-criteria decision support system to determine the suitable areas for wind turbine installation in the western region of Iran using GIS tools. They determined the wind potential of the suitable sites in the study area, similar to this study. There was no approach for determining the most appropriate spatial interpolation method in both three studies. The determination of the interpolation method is a significant step in wind speed map production. Each interpolation method has its own characteristics and gives different results.

The main purpose of the study was to determine the total wind energy potential in the Balıkesir Province by adopting an appropriate approach according to the legal regulations in Turkey. The wind speed map of the Balıkesir Province at a height of 100 m was produced considering the most appropriate spatial interpolation method. Subsequently, the wind power potential of the Balıkesir Province at a height of 100 m was calculated by excluding legally prohibited areas and those unsuitable for wind generation. The novelty in this study was the creation a wind turbine layout model in accordance with the national regulation, and the production of an equation that gives the number of turbines that can be installed in a certain area. In this way, an approach for assessing the wind energy potential for large areas in Turkey was put forward.

2. Study Area

The Balikesir Province is located in the northwest of Turkey. Some of the lands of the Balıkesir Province are in the Marmara Region and some are in the Aegean Region. The province has a coast facing the Marmara Sea to the north and the Aegean Sea to the west. It also borders the Greek island of Lesbos, located in the Aegean Sea.

The Balıkesir Province has a significant role in wind energy production in terms of both its installed capacity and potential within Turkey. When the Balıkesir Province is considered in terms of its total installed power and power potential, it has great importance in terms of wind energy. The total wind energy potential at a height of 50 m in the Balıkesir Province is 13,827.36 MW, according to the wind energy potential atlas of Turkey (REPA) [43]. The total installed power amount was 1375.05 MW as of January 2022 [14]. It constitutes only 9.9% of the potential at a height of 50 m. Considering its installed capacity and new installations, it is seen that Balıkesir has developed in harmony with Turkey. The increase in turbine height due to developments in wind turbine technology provides an increase in wind energy potential. This means that the wind potential of the Balıkesir Province may be higher than that expressed in REPA. This supports the importance of determining the wind energy potential in the Balıkesir Province. Due to the current installed capacity, growth rate, and growth potential in Balıkesir, this province was determined as the case study area. The location of the Balıkesir Province in Turkey and a physical map are given in Figure 2.

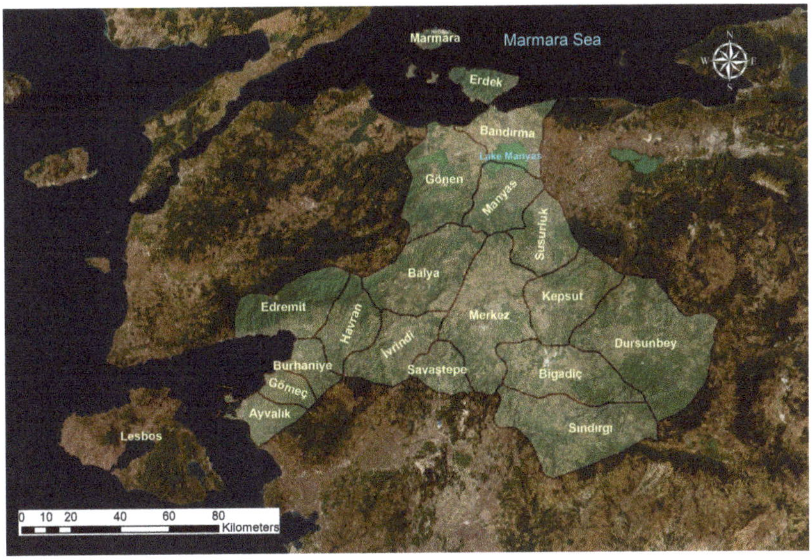

Figure 2. The location of Balıkesir Province and its districts.

3. Materials and Methods

The first step of the study was acquisition of the wind speed data belonging to meteorological stations and land cover data, in order to extrapolate the wind speeds to 100 m. Land cover data were used to determine the friction coefficient according to the roughness of the surface at the station. After the data acquisition, the wind speeds were extrapolated to 100 m by using the Hellmann equation. Following the calculation of the wind speeds of all the stations at a height of 100 m, the wind speed map of the study area was created with five different spatial interpolation methods of GIS tools. Interpolation methods were examined in terms of their accuracy and the most appropriate method was determined. Wind speed maps of the districts were produced by using the extract by mask function of GIS tools. The wind speed maps were converted to point-based shapefile by using the raster to point function of GIS tools to make wind potential calculations using the wind speed values of cells. The wind speeds in the study area were divided into certain ranges to determine the wind energy potential in a related range.

In order to calculate the number of wind turbines that can be installed in a certain area, an equation was produced according to the legal regulation in Turkey. Then, the power generated by one turbine was calculated according to the mean wind speed of each wind speed range. Then, the wind energy potential was calculated for each wind speed range. Finally, the total wind energy potential in the study area was obtained. Database files (.dbf) of the point-based wind speed maps were used to calculate the wind potential values. Calculations were made by using the Microsoft excel program. Details of the methodology are given below. Figure 3 shows the methodological framework of the study. Descriptions and sources of the data are shown in Table 3.

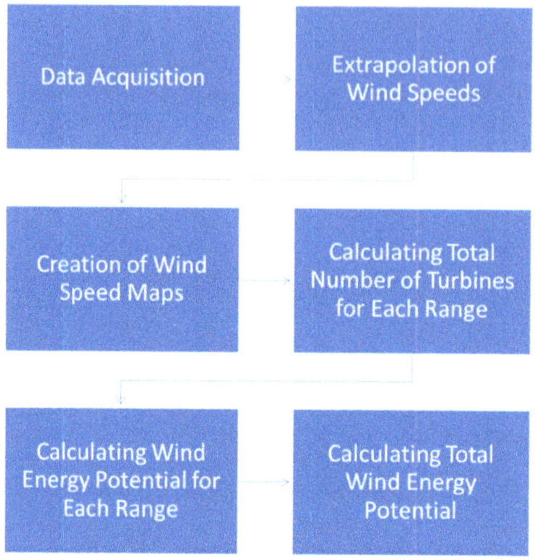

Figure 3. Methodological framework of the study.

Table 3. Data sources.

Data	Content	Source
Wind speed	Wind speed at a height of 10 m	Turkish State Meteorological Service
Corine land cover	Land cover types	European Environment Agency (EEA) [44]
ASTGTM v003	Digital elevation model	The United States Geological Survey (USGS) Earth Explorer [45]
Regional environmental plan	Residential areas, tourism areas, airports, areas of nature reserves, military areas, mining areas, organized industrial regions, logistics centers, small industrial areas, organized agriculture and livestock regions, work areas such as industry and storage areas, and water surfaces	General Directorate of Spatial Planning
GADM boundary map	Province and district boundaries	Global Administrative Areas

Within the scope of the study, 5.5 years of average daily wind speed data from 32 meteorology stations in the Balıkesir Province between 1 January 2014 and 30 June 2019 were obtained from the Turkish State Meteorological Service (TSMS). The geographical distribution of the TSMS stations used in the study is shown in Figure 4. The wind speed measurements at these TSMS stations are carried out on poles with a height of 10 m. Therefore, it is more meaningful to determine wind speeds according to a turbine hub height suitable for today's technology, in order to produce supportive information about possible wind power plant locations.

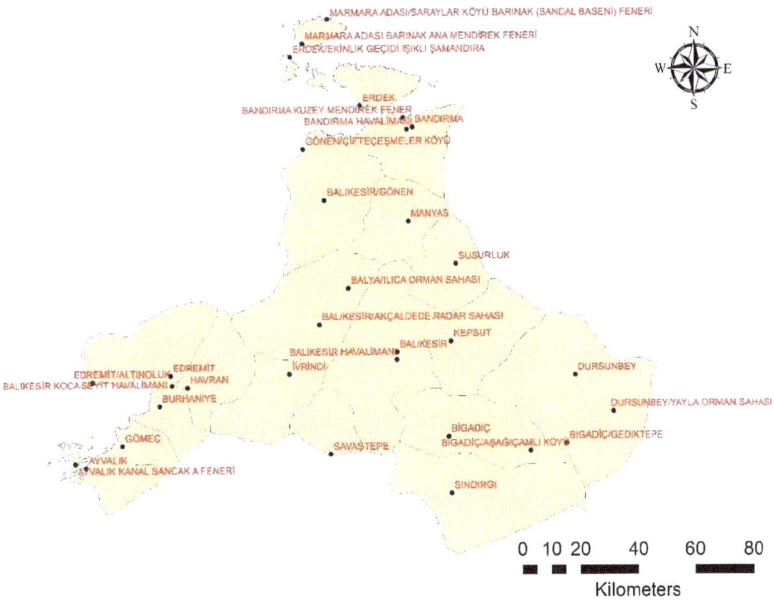

Figure 4. The geographical distribution of TSMS stations.

In this regard, the Hellmann equation shown in Equation (1) was used to obtain the wind speeds at a 100 m height from the measured wind speed values at a 10 m height. In this approach, calculations are made depending on the surface roughness of the region where the TSMS stations are located and the height at which the speed is required [46].

$$\frac{V}{V_0} = \left(\frac{H}{H_0}\right)^{\alpha} \tag{1}$$

In Equation (1), V_0 is the wind speed value for the TSMS station at a 10 m height, H is the height at which the wind speed is desired, α is the friction coefficient determined according to the roughness of the surface at the station where the wind is measured, and V is the wind speed value at the height desired. H_0 was taken as 10 m, at which the TSMS wind speeds were measured. Table 4 shows the land use characteristics and friction coefficients used in the study [47].

Table 4. Friction coefficient for different land use characteristics [47].

Land Use Characteristic	Friction Coefficient (α)
Smooth hard ground, calm water	0.10
Tall grass on level ground	0.15
High crops, hedges, and shrubs	0.20
Wooded countryside, many trees	0.25
Small town with trees and shrubs	0.30
Large city with tall buildings	0.40

A CORINE land cover map was obtained from the Copernicus Services inventory to obtain the friction coefficients for the locations of the TSMS stations [44]. The land cover class for the locations of the stations was matched with the appropriate class among the roughness classes used in the wind speed calculations, and the speed values of that station at a height of 100 m were calculated with the help of Equation (1). The reason for choosing a height of 100 m here is that the height of the turbine hubs in currently installed wind power plants is close to this value. In this way, the total installed wind power capacity and the calculated wind power potential can be compared more meaningfully.

After calculating the wind speeds at a 100 m height, interpolation between 32 TSMS stations was carried out with the inverse distance weighting (IDW) method to create a wind speed map of the Balıkesir Province using the ArcMap software. The IDW interpolation method is based on the principle that the weights decrease as the area being interpolated gets further from the known sample points. ArcMap is a GIS software that has the capability to collect, store, analyze, and present geo-spatial, attribute, and metadata. According to Yildiz [48], the IDW method was determined as the most appropriate interpolation method compared to the kriging, natural neighbor, spline, and trend methods. The measured wind speed values of the TSMS stations were compared with the values obtained using the interpolation methods. Root mean square error (RMSE) values obtained from the comparison of the measured wind speed values and the values obtained using the interpolation method were calculated for each interpolation method. The RMSE value of the IDW method was calculated as 0.008 m/s, which was the minimum RMSE value between these five methods. When the RMSE values of the other interpolation methods were examined, the order from the lowest to highest RMSE values appeared as the natural neighbor, spline, kriging, and trend methods [48].

The wind speed map of the Balıkesir Province at a height of 100 m was produced with 200×200 m cells using the IDW method under the raster interpolation tool in the ArcMap software, which is a GIS software (Figure 5). In addition, the extract by mask function of GIS tools was performed to produce wind speed maps of the districts in the Balıkesir Province. All of the produced wind speed maps are in raster data format.

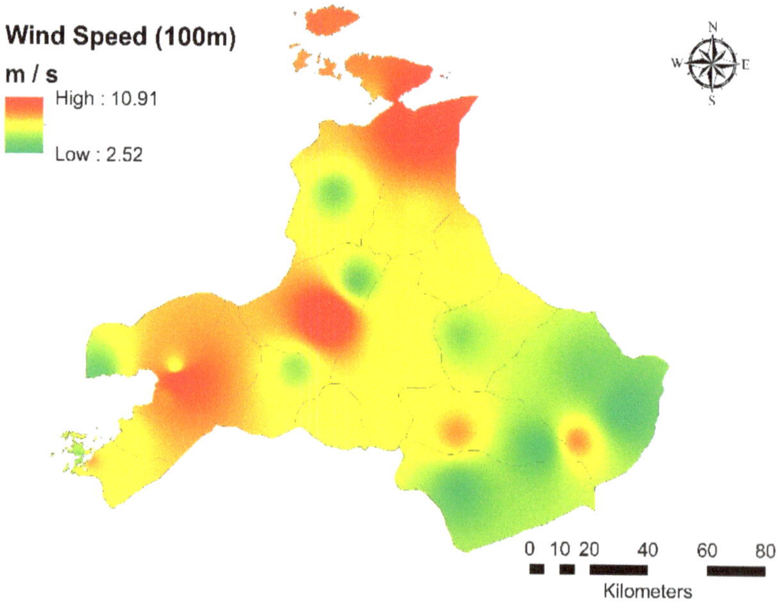

Figure 5. Wind speed map of Balıkesir Province at a height of 100 m.

The number of wind turbines that can be installed in a certain area was determined to calculate the wind energy potential at a height of 100 m. In this process, an equation was produced that gives the number of turbines that can be installed in a certain area, taking into account two rows side by side and a layout plan along the route, depending on the rules of the Regulation on the Technical Evaluation of Wind Source Based Electricity Generation Applications in Turkey [49]. In this calculation, the turbine rotor diameter was taken as 120 m. The distance between the turbines was considered to be 840 m, which is seven times the rotor diameter along the route, and 360 m, which is three times the rotor diameter for the second row perpendicular to the route (Figure 6). Equation (2) refers to the number of turbines that can be placed in a certain area. In order to calculate the number of turbines, the amount of the total area where the wind turbines will be installed is needed. The total number of turbines that can be installed in a certain area is calculated by multiplying the total area with the constant value of 6.6138 produced according to the layout plan created in accordance with the national regulation and adding 2 to it.

$$\text{Number of Turbines} = 6.6138 \times \text{Area}\left(\text{km}^2\right) + 2 \tag{2}$$

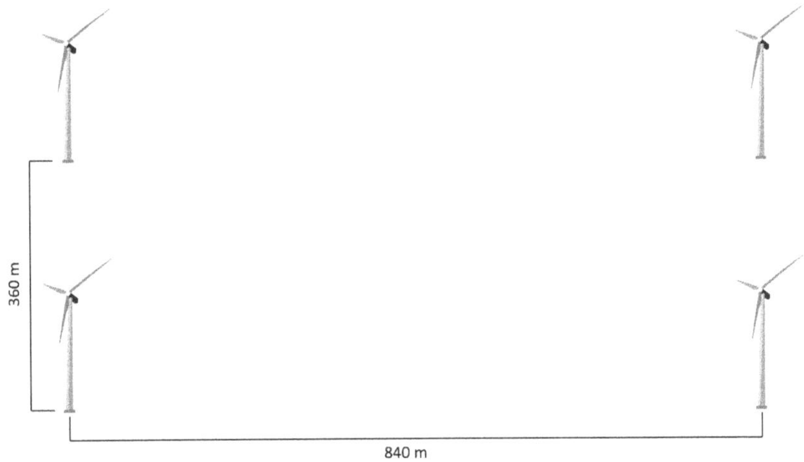

Figure 6. Sample siting plan of wind turbines.

In order to determine the wind energy potential in a certain area, it is necessary to calculate the power that each turbine can produce, as well as the number of turbines. Equation (3) was used for this calculation.

$$P = \frac{1}{2} \cdot \rho \cdot V^3 \cdot C_p \cdot A \tag{3}$$

In Equation (3), P is the power (Watt) generated by the turbine, ρ is the air density (equal to 1.225 kg/m^3), V is the mean wind speed for each range (m/s), C_p is the power coefficient (equal to 0.40 in general), and A is the area swept by the turbine blades (m^2).

In order to obtain the wind speed value of each cell to be used in the wind potential calculation, a conversion operation was carried out on the wind speed maps using the raster to point tool of the ArcMap software.

In order to determine the area-based wind energy potential, the wind speed values should be divided into certain ranges. The total area of the cells in each range was calculated by multiplying the number of cells in the specified range with the area of each cell. The wind power potential for each range was calculated by multiplying the number of wind turbines that can be installed in the total area for each range and the power generated by each wind turbine.

Spatial data obtained from the Republic of Turkey, Ministry of Environment and Urbanization, General Directorate of Spatial Planning were used to determine the exclusion areas. These areas are defined by the Republic of Turkey Energy Market Regulatory Authority (EPDK) as unsuitable areas for wind energy production in REPA due to legal and physical restrictions [43]. Residential areas, tourism areas, airports, areas of nature reserves, military areas, mining areas, organized industrial regions, logistics centers, small industrial areas, organized agriculture and livestock regions, work areas such as industry and storage areas, water surfaces, and areas where the slope is more than 30% were excluded from the study. A digital elevation model (ASTGTM v003) was obtained from the United States Geological Survey (USGS) Earth Explorer site in raster format [45]. A slope map was created from the digital elevation model by using the slope function of the ArcMap software. The excluded areas are shown in Figure 7. Since the reference ellipsoid of the TSMS station locations is the World Geodetic System 1984 (WGS 84), the WGS 84 ellipsoid was used as the reference ellipsoid in the study. Accordingly, all the geographical data obtained from institutions, such as landscaping plans, the digital elevation model, and administrative boundaries, were converted to the WGS 84 ellipsoid.

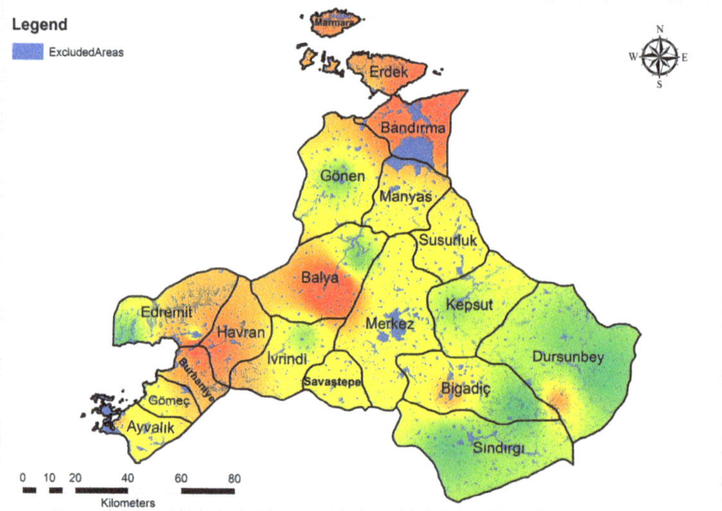

Figure 7. Excluded areas.

4. Results

When the wind speed map at a height of 100 m produced for the Balıkesir Province (Figure 5) is examined, the speeds are distributed between 2.52 and 10.91 m/s. When the speed distributions are evaluated for the districts, the high-speed values in the Balya, Bandırma, Edremit, Burhaniye, Havran, and Erdek districts are greater than those in the other districts. In addition, the maximum, minimum, and average wind speeds, and standard deviations at a height of 100 m are shown according to district in Table 5. These wind speeds were derived from the wind speed maps in raster data format. Figure 8 illustrates the wind speed maps of the Ayvalık, Balya, Bandırma, and Bigadiç districts at a height of 100 m. The wind speed maps of the other districts are also given in Figures A1–A4 of Appendix A.

Wind speed varies depending on landforms. While the friction effect of wind increases in mountainous and rough terrains, its speed decreases. On smooth surfaces, such as the sea and oceans, the wind speed increases as the friction effect decreases.

When the wind speed map created is examined, it is seen that the wind speeds are generally higher on the coastline. The Ayvalık, Gömeç, Burhaniye, Edremit, Gönen, Bandırma, Erdek, and Marmara districts are located on the coastline. To the west of the Ayvalık and Gömeç districts, there is the island of Lesbos. Due to the roughness caused by Lesbos Island, the wind speeds in the Ayvalık and Gömeç districts are lower than those in other districts on the coastline. In addition, the wind speeds in the İvrindi and Savaştepe districts are lower than those in other districts for the same reason. However, due to the air corridor formed by the passage of the Bosphorus between the Çanakkale Province and

Lesbos Island, the wind speeds are at high levels in the districts of Burhaniye, Edremit, Havran, and Balya, which are located in this direction. Since Sındırgı, Bigadiç, Dursunbey, and Kepsut are the furthest districts from the coastline, their wind speeds are lower than those of other districts.

Table 5. The maximum, minimum, average wind speeds, and standard deviations in the districts at a height of 100 m.

Name of District	Wind Speed (m/s)			
	Max	Min	Mean	Std. Dev.
Ayvalık	6.79	4.21	6.00	0.37
Balya	10.91	3.79	6.73	1.44
Bandırma	10.82	6.63	8.17	0.67
Bigadiç	6.86	2.85	5.27	0.93
Burhaniye	10.02	6.16	7.12	0.55
Dursunbey	7.05	2.52	4.47	0.79
Edremit	10.05	3.56	6.16	0.88
Erdek	8.36	6.39	7.44	0.48
Gömeç	6.75	5.63	6.11	0.31
Gönen	7.59	4.03	5.80	0.68
Havran	9.58	6.01	6.91	0.43
İvrindi	7.20	4.51	5.86	0.59
Kepsut	5.77	4.30	5.13	0.29
Manyas	7.19	5.42	6.10	0.34
Marmara	7.49	6.82	7.27	0.19
Merkez (Altıeylül-Karesi)	7.34	4.75	5.71	0.31
Savaştepe	5.94	5.36	5.64	0.13
Sındırgı	5.91	3.07	4.53	0.55
Susurluk	6.40	5.16	5.89	0.21

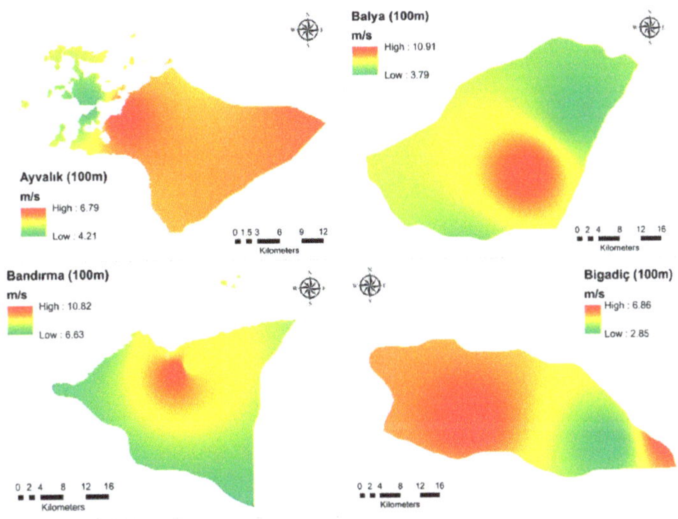

Figure 8. Wind speed maps of Ayvalık, Balya, Bandırma, and Bigadiç districts at a height of 100 m.

When the maximum and average wind speeds and standard deviations are examined, it is seen that Bandırma is the most favorable district in terms of wind speed. There is the Marmara Sea to the north of Bandırma and Lake Manyas to the south. Therefore, the roughness that will prevent the wind flow in the district is low. For this reason, wind speeds are at high levels throughout the district.

Since wind speed is the main factor in the calculation of wind power potential, the distribution of wind power potential in the districts also differs due to the reasons stated above and is parallel to the wind speed distribution.

Moreover, the wind energy potentials of the districts were also calculated to compare between the districts and give efficient information to decision makers. Table 6 demonstrates the wind energy potential of each district, according to different wind speed threshold values. In a study conducted by the Turkish Atomic Energy Agency on alternative energy sources, the lowest wind speed that can be used for electricity generation was determined to be 6 m/s [50]. According to the International Electrotechnical Commission (IEC) standard, the minimum technical wind speed is expressed as 6 m/s by Zahedi et al. [42]. For these reasons, the minimum speed was determined to be 6 m/s in this study. Other threshold values were determined in order to compare the wind energy potentials of the districts.

Table 6. The wind energy potential of districts.

Name of District	Wind Energy Potential (MW)		
	>6 m/s	>7 m/s	>8 m/s
Ayvalık	548.49	0.00	0.00
Balya	5758.53	3780.93	2416.78
Bandırma	4957.45	4839.97	3668.77
Bigadiç	848.04	0.00	0.00
Burhaniye	1625.17	929.52	265.30
Dursunbey	334.24	23.90	0.00
Edremit	2031.30	236.08	83.45
Erdek	1639.09	1466.05	427.49
Gömeç	674.11	0.00	0.00
Gönen	2010.05	275.84	0.00
Havran	2986.36	1451.84	51.75
İvrindi	1294.61	56.03	0.00
Kepsut	0.00	0.00	0.00
Manyas	1679.05	0.00	0.00
Marmara	664.25	553.82	0.00
Merkez (Altıeylül-Karesi)	875.36	64.96	0.00
Savaştepe	0.00	0.00	0.00
Sındırgı	0.00	0.00	0.00
Susurluk	1101.49	0.00	0.00

The wind speed values were divided into certain ranges and the number of cells in the specified ranges and corresponding total areas were determined in order to more clearly understand the wind energy potential of the Balıkesir Province at a height of 100 m (Table 7). The wind energy potential values in Tables 6 and 7 were derived from the calculation process carried out by using the database files of the point-based shapefile of the wind speed maps.

Table 7. The wind energy potential of Balıkesir Province for specified ranges.

Wind Speed Range (m/s)	Number of Cells	Total Area (km^2)	Wind Potential (P) (MW)
2.5–4.5	42,794	1711.76	1345.22
4.5–6.0	157,450	6298.00	16,702.09
6.0–7.0	76,122	3044.88	15,325.79
7.0–7.5	14,768	590.72	4127.50
7.5–8.0	7616	304.64	2601.31
8.0–8.5	6645	265.80	2738.29
8.5–9.0	4067	162.68	2000.94
9.0–9.5	1483	59.32	823.39
9.5–10.0	784	31.36	537.81
>10	853	34.12	730.26

5. Discussion and Conclusions

With the approach proposed in this study, it is possible to calculate the wind energy potential for a certain area using multiple points, which is superior to the single point-based wind energy potential determination studies [28–39] mentioned in the introduction section. In these point-based studies, wind energy potential calculations can only be made for these points and areal results cannot be produced. With the spatial interpolation method, which is a GIS function, wind speeds for an entire area can be generated by interpolating the wind speeds belonging to multiple points in a certain area. In this way, the total wind energy potential can be calculated for a study area using wind speeds. The wind speed maps of the Balıkesir Province and its districts were produced using IDW, which is the most appropriate spatial interpolation method in this study. Area-based wind potentials of the study region were obtained for different ranges considering the wind turbine layout plan designed according to the national regulation.

Anwer and Deshmukh [40] used the kriging interpolation method to obtain a wind speed map of the southern states of India. The produced wind speed map was insufficient in terms of cartographic representation. Another missing point in the study is that no comparison was made between interpolation methods. Likewise, Feng et al. [41] also adopted the IDW interpolation method to generate a wind speed map of mainland China without comparing interpolation methods. The wind energy potential of the study area was determined by adopting the wind farm layout approach of Gustavson [51] using wind potential calculation algorithms. These calculations were made according to the specifications of a specific wind turbine. Turbine specifications were used to obtain the maximum power coefficient in the study. Zahedi et al. [42] used a GIS-based multi-criteria decision support system to determine the suitable areas for wind turbine installation in the western region of Iran. They used raw wind speed and roughness information via Wind atlas analysis and application program (WAsP) software and obtained a digital wind atlas. There was no information about interpolation methods and the process for creating the wind speed atlas in the study. A layout plan for the wind turbine installation was defined and the total wind energy potential was calculated for the study area. The number of wind turbines was given in the study, but no equations related to it were given. The total wind potential for a given wind turbine was determined.

In this study, raw wind speeds gathered from TSMS stations were extrapolated to 100 m considering the roughness of the terrain, and wind speed maps were created using the IDW, kriging, natural neighbor, spline, and trend interpolation methods. The most appropriate method was determined to be IDW with the minimum RMSE value of 0.008 m/s. In addition, a layout plan for wind turbine installation was designed considering the minimum distances required between two turbines in the Regulation on the Technical Evaluation of Wind Source Based Electricity Generation Applications, and an equation (Equation (2)) was produced that gives the number of wind turbines that can be installed in a certain area in accordance with this layout plan. Unlike other studies carried out, no existing turbine model was taken as a reference, so that the model did not become dependent on any brand. Wind energy investors can apply this model with the parameters of the turbine model they want.

Wind speeds and Equations (2) and (3) were used to calculate the wind energy potential for each wind speed range. Finally, the total wind energy potentials of the Balıkesir Province and its districts were obtained by summing the wind energy potentials of all the ranges.

When the threshold value for wind speed is taken as 6 m/s, the total wind power potential is calculated as 28,885.30 MW in the Balıkesir Province. It is calculated as 13,559.51 MW with a 7 m/s threshold value and 6830.69 MW with an 8 m/s threshold value.

The wind speed threshold value was taken as 6.8 m/s in the wind energy potential atlas of Turkey (REPA), and the total wind energy potential at a height of 50 m for the Balıkesir Province was determined to be 13,827.36 MW. In this study, the wind energy potential was calculated as 15,928.81 MW with a 6.8 m/s wind speed threshold. There are five different wind speed ranges in REPA. The wind energy potential for a given area is calculated by multiplying the total area value by five and is a general assumption. Although the total area value obtained in this study was lower than that in REPA, the calculated total wind power potential value was higher. This is because a realistic approach was used in this study to calculate the wind energy potential. The reason for the lower area values in this study is that areas were excluded from the study if they were legally prohibited or unsuitable areas. Another reason why the wind potential values were higher than those in REPA is that the height used in this study was greater than that used in REPA. The total area and wind potential values obtained in this study and REPA are represented in Table 8.

Table 8. Comparison of current study with REPA.

Wind Speed Range (m/s)	Current Study		REPA	
	Total Area (km^2)	Wind Potential (P) (MW)	Total Area (km^2)	Wind Potential (P) (MW)
6.8–7.5	965.28	6468.10	1511.42	7557.12
7.5–8.1	364.64	3173.79	850.96	4254.80
8.1–8.6	252.88	2701.24	284.51	1422.56
8.6–9.5	174.92	2380.15	115.23	576.16
>9.5	65.48	1205.53	3.34	16.72
Total	1823.20	15,928.81	2765.47	13,827.36

When the wind power potentials of the districts are examined according to the 6 m/s wind speed threshold value, the district with the highest potential is Balya. Bandırma, Havran, and Edremit follow Balya. If the threshold value is taken as 7 m/s, this order is Bandırma, Balya, Erdek, and Havran. When the threshold value is taken as 8 m/s, this order changes to Bandırma, Balya, Erdek, and Burhaniye. This change in the rankings is due to the wind speed distribution in the districts. When the wind power potential values in the districts are examined, it is obvious that Bandırma and Balya are much more efficient than the other districts.

The total installed power amount was 1375.05 MW as of January 2022. This value constitutes only 8.6% of the 15,928.81 MW wind energy potential value calculated as a result of this study. This shows that the Balıkesir Province is still quite suitable for wind energy investments. This study is expected to contribute to the evaluations of investors and decision makers working in the field of renewable energy.

The limitation of the method is that the wind energy potential value obtained within a certain area cannot be visualized. In order to visualize the results obtained, the energy potential of each cell must be calculated. The amount of error will increase when Equation (2), which calculates the number of wind turbines that can be installed in a certain area, is used for each cell in high-spatial-resolution studies. For example, the spatial resolution of this study is 200 m and the size of each cell is 200 × 200 m. According to the Regulation on the Technical Evaluation of Wind Source Based Electricity Generation Applications mentioned in Section 3, the distance between two turbines has to be at least 840 m, which is seven times the rotor diameter along the route, and 360 m, which is three times the rotor diameter for the second row perpendicular to the route. In this case, it is impossible to install more than one turbine in each cell. However, using Equation (2), the number of turbines that can be installed in each cell is calculated as 2.26. For this reason, in this study, the wind energy potential is calculated for each wind speed range by taking into account the total number of cells in the specified ranges, and these results cannot be visualized. On the other hand, in order to compare the obtained results with REPA, the same wind speed ranges were created with REPA and comparisons were made. In addition, studies will continue to calculate the wind energy potential that can be produced in each cell to visualize the results produced. The proposed methodology can be applied in all areas that consist of cells having an individual wind speed value.

The turbine rotor diameter used in the study was evaluated as 120 m and the hub height was evaluated as 100 m by considering the installed wind turbines in Turkey and up-to-date wind turbine technology. In order to not adopt a brand- and model-dependent approach, no turbine model of any brand was taken as a reference. The approach used in this study can be applied considering different hub heights and predetermined turbine specifications. Wood [52] mentioned problems that wind farm owners may encounter technically, such as a miscalculation of the power coefficient for the turbine, which may cause maintenance costs in the future. Therefore, more appropriate results can be obtained if the wind energy potential determination approach in this study is re-applied in projects to be realized in smaller areas according to the type of turbine determined, by taking into account meteorological conditions.

Funding: This research received no external funding.

Institutional Review Board Statement: Not applicable.

Informed Consent Statement: Not applicable.

Data Availability Statement: The results presented in this paper are available upon request. Corine Land Cover (CLC) 2018 data is public available and can be sourced at https://land.copernicus.eu/pan-european/corine-land-cover/clc2018?tab=download (accessed on 11 February 2023).

Conflicts of Interest: The author declares no conflict of interest.

Appendix A

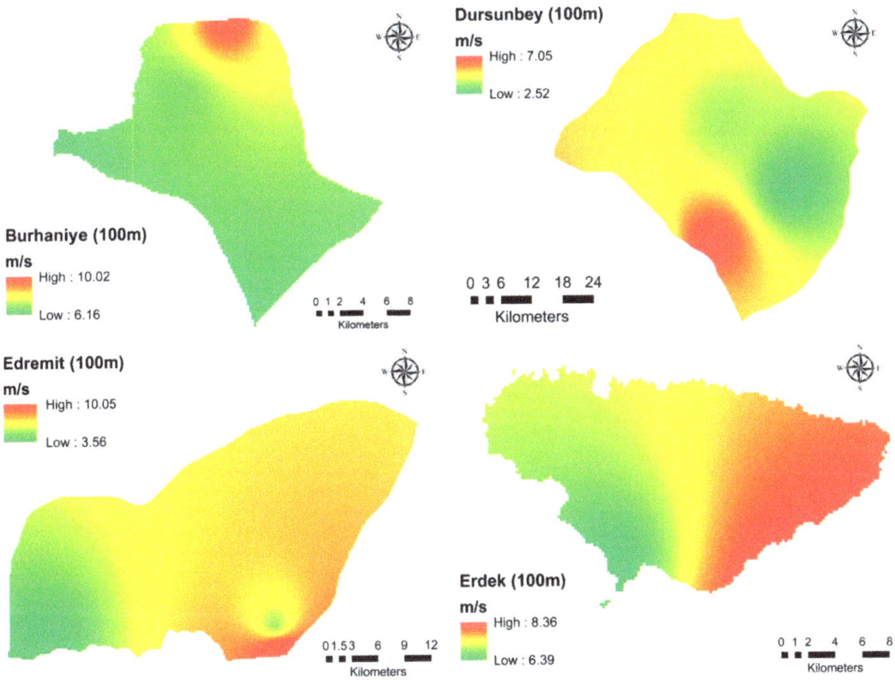

Figure A1. Wind speed maps of Burhaniye, Dursunbey, Edremit, and Erdek districts at a height of 100 m.

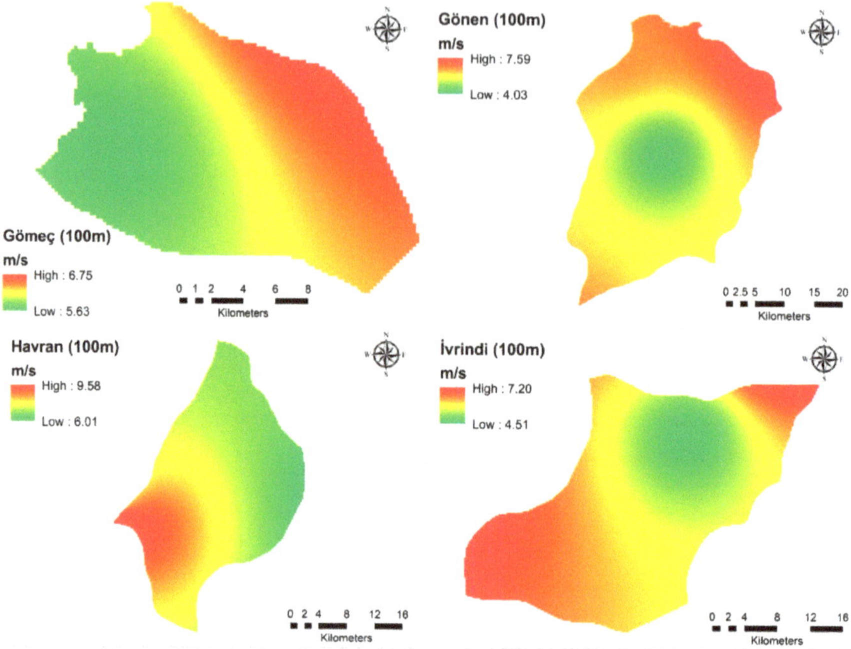

Figure A2. Wind speed maps of Gömeç, Gönen, Havran, and Ivrindi districts at a height of 100 m.

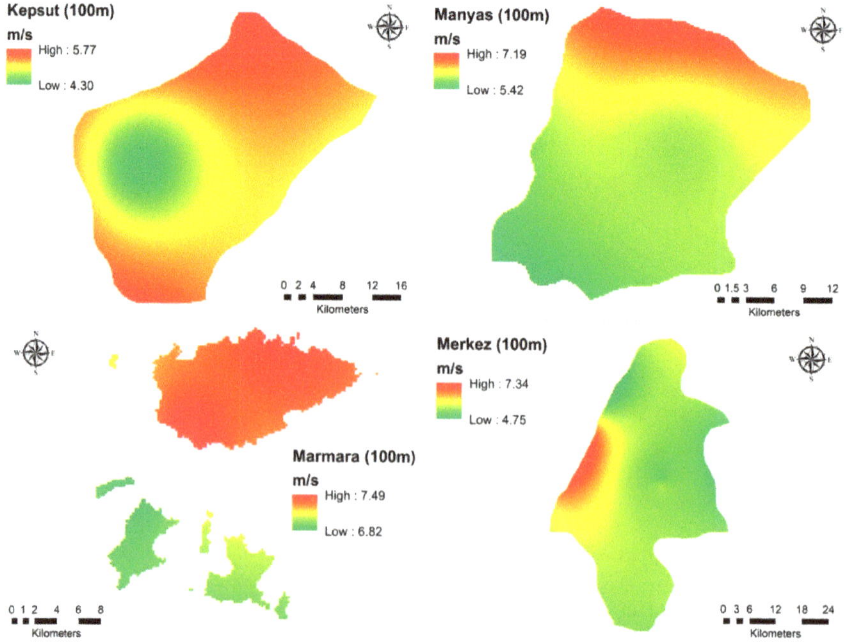

Figure A3. Wind speed maps of Kepsut, Manyas, Marmara, and Merkez (Altıeylül and Karesi) districts at a height of 100 m.

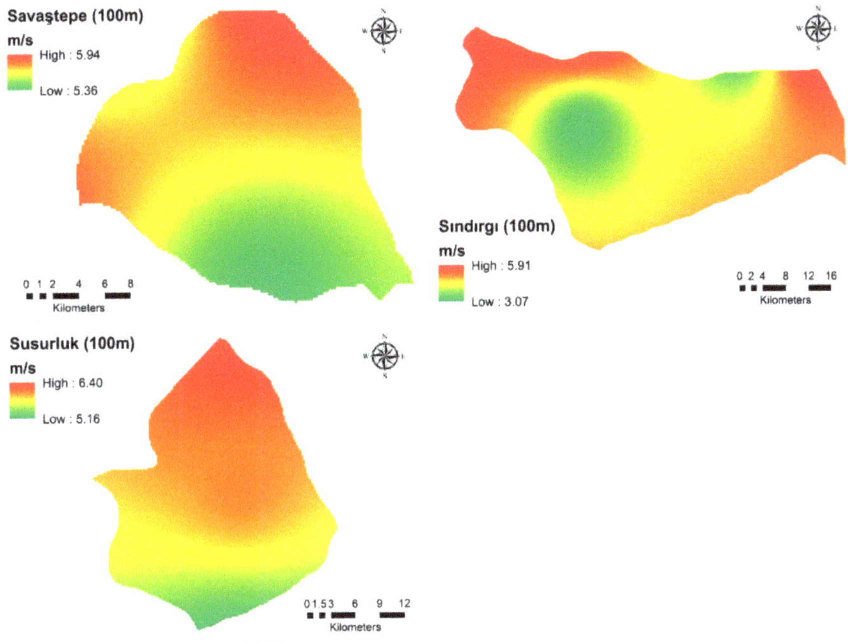

Figure A4. Wind speed maps of Savaştepe, Sındırgı, and Susurluk districts at a height of 100 m.

References

1. Marks-Bielska, R.; Bielski, S.; Pik, K.; Kurowska, K. The Importance of Renewable Energy Sources in Poland's Energy Mix. *Energies* **2020**, *13*, 4624. [CrossRef]
2. Piwowar, A.; Dzikuć, M. Development of renewable energy sources in the context of threats resulting from low-altitude emissions in rural areas in poland: A review. *Energies* **2019**, *12*, 3558. [CrossRef]
3. Erdin, C.; Ozkaya, G. Turkey's 2023 energy strategies and investment opportunities for renewable energy sources: Site selection based on electre. *Sustainability* **2019**, *11*, 2136. [CrossRef]
4. Shahzad, U. The need for renewable energy sources. *Int. J. Inf. Technol. Electr. Eng.* **2015**, *4*, 16–18.
5. Dincer, F. The analysis on wind energy electricity generation status, potential and policies in the world. *Renew. Sustain. Energy Rev.* **2011**, *15*, 5135–5142. [CrossRef]
6. Lee, J.; Zhao, F. *GWEC Global Wind Report 2021*; Global Wind Energy Council (GWEC): Brussels, Belgium, 2021.
7. Lee, J.; Zhao, F. *GWEC Global Wind Report 2022*; Global Wind Energy Council (GWEC): Brussels, Belgium, 2022.
8. Global Wind Power Statistics by 2021. Available online: https://library.wwindea.org/global-wind-power-statistics/ (accessed on 20 March 2023).
9. Dawn, S.; Tiwari, P.K.; Goswami, A.K.; Singh, A.K.; Panda, R. Wind power: Existing status, achievements and government's initiative towards renewable power dominating India. *Energy Strategy Rev.* **2019**, *23*, 178–199. [CrossRef]
10. Presidency of the Republic of Turkey Presidency of Strategy and Budget: Republic of Turkey's Eleventh Development Plan Covering the 2019–2023 Period. Available online: https://www.sbb.gov.tr/wp-content/uploads/2022/07/Eleventh_Development_Plan_2019-2023.pdf (accessed on 13 February 2023).
11. General Directorate of Energy Affairs: Turkey Electricity Investments in June 2022 Summary Report. Available online: https://enerji.gov.tr//Media/Dizin/EIGM/tr/Raporlar/PRP/EY%C3%96RA/2022/Haziran/ElektrikYat%C4%B1r%C4%B1mlar%C4%B12022HaziranAy%C4%B1%C3%96zetRaporu.pdf (accessed on 13 February 2023).
12. Gultekin, U. Development of wind energy investments in Turkey. *J. Turk. Stud.* **2019**, *14*, 2333–2348. [CrossRef]
13. Akdoğan, İ.; Kovancılar, B. Evaluation of Eco-Friendly Renewable Energy Policies in The European Union and Turkey in Terms of Incentive Types. *Yönetim ve Ekonomi Dergisi* **2022**, *29*, 69–91. [CrossRef]
14. Turkish Wind Energy Association. Turkish Wind Energy Statictic Report—January 2022. Available online: https://tureb.com.tr//yayinlar/turkiye-ruzgar-enerjisi-istatistik-raporlari/5 (accessed on 17 March 2023).
15. Turkish Wind Energy Association. Turkish Wind Energy Statictic Report—January 2021. Available online: https://tureb.com.tr//yayin/turkiye-ruzgar-enerjisi-istatistik-raporu-ocak-2021/139 (accessed on 17 March 2023).

16. Manfren, M.; Nastasi, B.; Groppi, D.; Astiaso Garcia, D. Open data and energy analytics—An analysis of essential information for energy system planning, design and operation. *Energy* **2020**, *213*, 118803. [CrossRef]
17. GoldWind Gw175-8.0MW Smart PMDD Wind Turbine Technical Parameters. Available online: https://www.goldwind.com/en/windpower/product-gw6s/ (accessed on 13 March 2023).
18. Vestas V172-7.2 MW™ Wind Turbine Technical Specifications. Available online: https://www.vestas.com/en/products/enventus-platform/V172-7-2-MW (accessed on 13 March 2023).
19. Nordex N163/6.X Wind Turbine Technical Data. Available online: https://www.nordex-online.com/en/product/n163-6-x/ (accessed on 13 March 2023).
20. Voivontas, D.; Assimacopoulos, D.; Mourelatos, A.; Corominas, J. Evaluation of Renewable Energy Potential Using A GIS Decision Support System. *Renew. Energy* **1998**, *13*, 333–344. [CrossRef]
21. Rodman, L.C.; Meentemeyer, R.K. A geographic analysis of wind turbine placement in Northern California. *Energy Policy* **2006**, *34*, 2137–2149. [CrossRef]
22. Al-Yahyai, S.; Charabi, Y.; Gastli, A.; Al-Badi, A. Wind farm land suitability indexing using multi-criteria analysis. *Renew. Energy* **2012**, *44*, 80–87. [CrossRef]
23. Gorsevski, P.V.; Cathcart, S.C.; Mirzaei, G.; Jamali, M.M.; Ye, X.; Gomezdelcampo, E. A group-based spatial decision support system for wind farm site selection in Northwest Ohio. *Energy Policy* **2013**, *55*, 374–385. [CrossRef]
24. Höfer, T.; Sunak, Y.; Siddique, H.; Madlener, R. Wind farm siting using a spatial Analytic Hierarchy Process approach: A case study of the Städteregion Aachen. *Appl. Energy* **2016**, *163*, 222–243. [CrossRef]
25. Noorollahi, Y.; Yousefi, H.; Mohammadi, M. Multi-criteria decision support system for wind farm site selection using GIS. *Sustain. Energy Technol. Assess.* **2016**, *13*, 38–50. [CrossRef]
26. Villacreses, G.; Gaona, G.; Martínez-Gómez, J.; Jijón, D.J. Wind farms suitability location using geographical information system (GIS), based on multi-criteria decision making (MCDM) methods: The case of continental Ecuador. *Renew. Energy* **2017**, *109*, 275–286. [CrossRef]
27. Baseer, M.A.; Rehman, S.; Meyer, J.P.; Alam, M.M. GIS-based site suitability analysis for wind farm development in Saudi Arabia. *Energy* **2017**, *141*, 1166–1176. [CrossRef]
28. Kavak Akpinar, E.; Akpinar, S. Determination of the wind energy potential for Maden-Elazig, Turkey. *Energy Convers. Manag.* **2004**, *45*, 2901–2914. [CrossRef]
29. Keyhani, A.; Ghasemi-Varnamkhasti, M.; Khanali, M.; Abbaszadeh, R. An assessment of wind energy potential as a power generation source in the capital of Iran, Tehran. *Energy* **2010**, *35*, 188–201. [CrossRef]
30. Fyrippis, I.; Axaopoulos, P.J.; Panayiotou, G. Wind energy potential assessment in Naxos Island, Greece. *Appl. Energy* **2010**, *87*, 577–586. [CrossRef]
31. Ohunakin, O.S.; Akinnawonu, O.O. Assessment of wind energy potential and the economics of wind power generation in Jos, Plateau State, Nigeria. *Energy Sustain. Dev.* **2012**, *16*, 78–83. [CrossRef]
32. Wu, J.; Wang, J.; Chi, D. Wind energy potential assessment for the site of Inner Mongolia in China. *Renew. Sustain. Energy Rev.* **2013**, *21*, 215–228. [CrossRef]
33. Elnaggar, M.; Edwan, E.; Ritter, M. Wind Energy Potential of Gaza Using Small Wind Turbines: A Feasibility Study. *Energies* **2017**, *10*, 1229. [CrossRef]
34. Elie Bertrand, K.S.; Abraham, K.; Lucien, M. Sustainable Energy Through Wind Speed and Power Density Analysis in Ambam, South Region of Cameroon. Front. *Energy Res.* **2020**, *8*, 176. [CrossRef]
35. Wang, Z.; Liu, W. Wind energy potential assessment based on wind speed, its direction and power data. *Sci. Rep.* **2021**, *11*, 16879. [CrossRef]
36. Çakmakçı, B.A.; Hüner, E. Evaluation of wind energy potential: A case study. *Energy Sources Part A Recovery Util. Environ. Eff.* **2022**, *44*, 834–852. [CrossRef]
37. Paraschiv, S.; Paraschiv, L.; Alexandru, S.; Anisoara-Gabriela, C. Assessment of onshore wind energy potential under temperate continental climate conditions. *Energy Rep.* **2022**, *8*, 251–258. [CrossRef]
38. Jung, C.; Schindler, D. Introducing a new approach for wind energy potential assessment under climate change at the wind turbine scale. *Energy Convers. Manag.* **2020**, *225*, 113425. [CrossRef]
39. Ongaki, L.; Maghanga, C.M.; Kerongo, J. Evaluation of the Technical Wind Energy Potential of Kisii Region Based on the Weibull and Rayleigh Distribution Models. *J. Energy* **2021**, *2021*, 6627509. [CrossRef]
40. Anwar, K.; Deshmukh, S. Parametric study for the prediction of wind energy potential over the southern part of India using neural network and geographic information system approach. *Proc. Inst. Mech. Eng. Part A J. Power Energy* **2020**, *234*, 96–109. [CrossRef]
41. Feng, J.; Feng, L.; Wang, J.; King, C.W. Evaluation of the onshore wind energy potential in mainland China—Based on GIS modeling and EROI analysis. *Resour. Conserv. Recycl.* **2020**, *152*, 104484. [CrossRef]
42. Zahedi, R.; Ghorbani, M.; Daneshgar, S.; Gitifar, S.; Qezelbigloo, S. Potential measurement of Iran's western regional wind energy using GIS. *J. Clean. Prod.* **2022**, *330*, 129883. [CrossRef]
43. Wind Energy Potential Atlas of Balıkesir, Turkey. Available online: https://repa.enerji.gov.tr/REPA/BALIKESIR-REPA.pdf (accessed on 16 February 2023).

44. Corine Land Cover (CLC). 2018. Available online: https://land.copernicus.eu/pan-european/corine-land-cover/clc2018?tab=download (accessed on 11 February 2023).
45. The United States Geological Survey (USGS). ASTGTMv003 ASTER Global Digital Elevation Model. Available online: https://lpdaac.usgs.gov/products/astgtmv003/ (accessed on 11 February 2023).
46. Tar, K. Some statistical characteristics of monthly average wind speed at various heights. *Renew. Sustain. Energy Rev.* **2008**, *12*, 1712–1724. [CrossRef]
47. Masters, G.M. *Renewable and Efficient Electric Power Systems*; John Wiley and Sons: Hoboken, NJ, USA, 2004.
48. Yildiz, S.S. A Comparison of Interpolation Methods in Creation of Wind Speed Maps: A Case Study of Balıkesir. *Afyon Kocatepe Univ. J. Sci. Eng.* **2021**, *21*, 130–137. [CrossRef]
49. Republic of Turkey Ministry of Energy and Natural Resources: Regulation on the Technical Evaluation of Wind Source Based Electricity Generation Applications. Available online: https://www.resmigazete.gov.tr/eskiler/2015/10/20151020-2-1.pdf (accessed on 17 February 2023).
50. Turkish Atomic Energy Agency. Alternatif Enerji Kaynakları. İstanbul: Çekmece Nükleer Araştırma ve Eğitim Merkezi. Available online: https://kurumsalarsiv.tenmak.gov.tr/bitstream/20.500.12878/1272/1/3308.pdf (accessed on 18 February 2023).
51. Gustavson, M.R. Limits to wind power utilization. *Science* **1979**, *204*, 13–17. [CrossRef]
52. Wood, D. Grand Challenges in Wind Energy Research. *Front. Energy Res.* **2020**, *8*, 624646. [CrossRef]

applied sciences

MDPI

Article

A Complete and High-Resolution Estimate of Sardinia's Rooftop Photovoltaic Potential

Andrea Pinna * **and Luca Massidda**

CRS4, Center for Advanced Studies, Research and Development in Sardinia, Loc. Piscina Manna ed. 1, 09050 Pula, CA, Italy

* Correspondence: andrea.pinna@crs4.it

Abstract: The implementation of the energy transition and the building of energy communities are driving forward the exploitation of the potential for rooftop photovoltaic power generation. Estimating rooftop PV generation potential requires the processing of different types of data, such as the cadastral information of buildings, a detailed description of available rooftop areas, and solar irradiance data. High-resolution estimation based on GIS data is normally limited to small survey areas. Instead, by using an algorithm for the efficient calculation of shadows over rooftops, and the integration of solar irradiance over time, we developed a procedure that allows for the rapid full census assessment of rooftop photovoltaic potential with a spatial resolution of 1 m, applicable at the regional scale and requiring minimal computational resources. We applied this approach to the rooftops of buildings in Sardinia, an island and region of Italy of particular interest for the energy transition. In addition to estimating the geographic potential, we carried out a preliminary assessment of the technical and economic potential, yielding a maximal photovoltaic rooftop generation potential of 22 TWh for the entire region.

Keywords: rooftop solar photovoltaic (PV) potential; geographic information systems (GIS); LiDAR; distributed generation

Citation: Pinna, A.; Massidda, L. A Complete and High-Resolution Estimate of Sardinia's Rooftop Photovoltaic Potential. *Appl. Sci.* **2023**, *13*, 7. https://doi.org/10.3390/app13010007

Academic Editors: Yannis Maniatis and Gaetano Zizzo

Received: 14 November 2022
Revised: 2 December 2022
Accepted: 16 December 2022
Published: 20 December 2022

1. Introduction

In recent years, the European Union has been very committed to increasing the amount of energy generated from renewable sources. The first target, set with the Renewable Energy Directive of 2009 [1], aimed at increasing the percentage of energy from renewable sources to 20% by 2020; the Renewable Energy Directive II of 2018 [2] set this percentage to 32%, to be reached by 2030. The REPowerEU directive from May 2022 [3] proposed to increase this target to 45% by 2030 in order for the European Union to become increasingly autonomous from importing fossil fuels.

In this paper, we deal with the generation of electricity from solar radiation by means of photovoltaic panels, which is particularly widespread in Italy (in 2021, it met 8.7% of the national electricity demand, and 21.8% of the production from renewable sources [4]). The installation of photovoltaic systems has been promoted by the Italian government, e.g., by way of incentives such as "Superbonus 110%" [5,6].

Building rooftops offer significant potential for the deployment of photovoltaic (PV) systems, and enable better geographic correlation between supply and demand [7]. The installation of photovoltaic modules has low impact from the point of view of urban planning, since the panels are mounted onto existing roofs.

Energy produced by photovoltaic systems can play a crucial role in reducing greenhouse gas emissions [8]; the contribution of rooftop photovoltaic systems is essential in smart cities [9], and the estimation of the production potential is a prerequisite for the transformation of cities and energy communities [10,11] into net zero energy districts [12].

The estimation of the photovoltaic potential is, therefore, required in various fields of study and research. The determination of the photovoltaic potential in urban areas consists of four basic parts [13,14]: physical potential, geographic potential, technical potential, and economic potential. (i) Physical potential consists of assessing the solar irradiation over the region of interest; (ii) geographic potential consists of identifying and characterizing the surfaces suitable for photovoltaic installations; (iii) technical potential studies the transformation of the solar energy harvested by the panels into electric energy; (iv) economic potential evaluates the return on the investment incurred in the installation of photovoltaic systems.

According to the literature, different techniques are used according to the scale of the analysis to be conducted [15]. Studies on a continental [7] or global [16] scale have a lower level of detail, with analyses primarily using statistical methods and machine learning techniques.

Joshi et al. carried out a global-scale estimation of the technical potential of rooftop PV systems [16]. Earth's surface is divided into sectors of 10 km^2, and the total area of building roofs is estimated for each with a machine-learning approach on the basis of population density, and geographic information system (GIS) datasets. The result is the technical and economic potentials calculated for all countries and for each month of the average year. Bodis et al., using satellite and statistical data, and machine-learning techniques, estimated the photovoltaic potential on EU rooftops at 100 m spatial resolution [7].

Statistical methods are still used in national- [17] and regional [18]-scale studies, often being supplemented with GIS data [19].

The use of machine-learning algorithms coupled with high-resolution GIS data and physical models allowed for Walch et al. to estimate with a high level of detail the large scale rooftop photovoltaic potential in Switzerland [19]. An evaluation of the geographic potential is realized by combining the estimated average monthly irradiance obtained from satellite data with available rooftop area, shading effects and rooftop geometry. Lastly, the use of physical models allows for the estimation of the technical potential and its uncertainty. The work of Walch et al. is a notable exception to other studies on the same geographic scale because of the level of detail in the analysis. Regional-scale analyses are typically carried out using statistical methods for extrapolating detailed analyses to samples. Bernasconi et al., for example, estimated the area available for PV installation using the clustering analysis of the urban and morphological characteristics of municipalities in Lombardy, the most densely populated region in Italy with the largest number of inhabitants [18].

Analyses on an urban scale, on the other hand, require the use of GIS data [20] or light detection and ranging (LiDAR) surveys [21,22], or both [23,24], and other data sources such as satellite images [25] or aerial photographic images [26] processed with machine-learning techniques [27–29].

In [20], for an area of the city of Turin (Italy) having an extension of 1 km^2 and containing 1228 buildings, GIS tools were used to estimate the tilt and orientation of rooftops from high-resolution (0.5 m) digital surface models. The problem of shading and effective irradiance had already been addressed in [23], where the sky view factor [30,31], and obstructions due to terrain and buildings were calculated with LiDAR data for a small area in the city of Kingston, Ontario (Canada), and in [21], in which a three-dimensional model of buildings was developed to calculate shading and thus the actual irradiance on rooftops in the municipality of Avellino (Italy).

More recently, in [22], a method was proposed for the automatic identification of roofs suitable for photovoltaic systems. The technique was demonstrated on buildings in the city of Gothenburg (Sweden) using a digital model of the surface with a resolution of 10 cm as the only input. The algorithm presented in [25] allowed for the examination of a 1 km^2 area at high resolution in the city of Giessen (Germany).

In [24], a methodology was developed for the estimation of the rooftop potential for food, water, and photovoltaics in the city of Zaragoza (Spain), where LiDAR and cadastral data were used to identify suitable covers for the different purposes. Recently, techniques in machine learning have been added to the GIS methodologies mentioned above, such as in [26] where a fast method was proposed to three-dimensionally construct the geometry of buildings in an area of the Netherlands through the use of stereo aerial imagery with 10 cm resolution and cadastral information, while deep-learning techniques have found use for the segmentation of suitable roof surfaces in urban areas [27,29].

Our work illustrates a complete methodology that allows for the high-resolution estimation of the photovoltaic potential at a regional scale. We applied the procedure on a large scale to an entire region of Italy, the island of Sardinia, and expanded the results presented at the city scale in [32]. Sardinia has great potential for generation from renewable sources and is included in the Coal Regions in Transition (CRiTs) initiative supported by the European Union [33]; therefore, it needs to develop and implement projects to realize a viable economic and technological transformation.

As far as physical potential is concerned, we reworked the data provided by PVGIS [34]; in terms of the technical potential, and calculated the average energy production by means of a simplified formulation [35]; the economic potential was evaluated by means of an estimation of the levelized cost of electricity (LCOE) [36]. Our method focuses on the geographic potential and thus uses open cadastral, GIS, and LiDAR data to estimate the available rooftop surface, to approximate the inclination and orientation of building rooftops, and to evaluate the shading effects between buildings [32].

The scientific contribution is related to the extremely effective algorithm used for time integration of solar irradiance reaching a surface, accounting for the shading effects of other buildings, which can be applied to the roof surfaces of buildings in the entire region with a spatial resolution of 1 m. Our procedure allows for us to combine a very high level of detail within the extent of the area under consideration. The proposed methodology, in terms of the employed types of techniques, the level of detail, and the extent of the examined area, can be directly compared with the work of Walch et al. [19]. However, in our case, we propose a more detailed analysis of the effects of shading and calculation of the average annual irradiance. In fact, due to the efficiency of our procedure, we were able to integrate over time the solar irradiance incident on each surface pixel modulated by shading effects with a temporal resolution of one hour, whereas in the cited work, the calculation was conducted on an hourly basis and with a monthly averaged irradiation.

By employing open data and open-source software, our procedure was able to estimate the technical potential for the entire region of Sardinia on a modest workstation and in a reasonable time (less than a week), thus demonstrating, contrary to common understanding [37], that it is possible to obtain a timely estimation of photovoltaic potential with a LiDAR-based approach, even for large areas, by employing limited computational resources.

Section 2 describes the data used in the study; Section 3 explains the methodologies used to process the data to obtain the physical, geographic, technological, and economic potential; Section 4 illustrates the results obtained on a regional level; Section 5 discusses the results; Section 6 draws conclusions and addresses possible future developments.

2. Data

2.1. Introduction

The energy that may be generated by a photovoltaic system can be estimated by knowing (i) the location of the system, (ii) the tilt and orientation of the panels, (iii) the parameters that characterize the modules and the inverter, (iv) the temperature, the wind speed, and the average amount of direct and diffuse radiation at the site, and (v) the presence of any shading that may affect the amount of direct radiation reaching the panels.

In the following paragraphs, we illustrate the procedures for acquiring and processing the data required to obtain our estimate, i.e., the building blueprints (Section 2.2), the elevation raster (Section 2.3), the solar radiation measurements, and the meteorological data (Section 2.4).

2.2. Building Blueprints

Knowledge of the correct position of a building allows for us to associate it with the average irradiance estimated for that geographic area, while the relative position with respect to other buildings and their elevation allow for us to calculate the possible shading that reduces the overall irradiance received.

The Geotopographic Database (DBGT) of Sardinia [38] provides the regional technical authoritative maps in digital format (DBGT10K, published in 2017). Among the various available layers, we were interested in the shape files representing the building blueprints (Figure 1) and the municipal boundaries.

In the same regional geoportal, the DBGT10K updated to 2020 is available, although only for the southern part of the island. For the purposes of this study, we used the building blueprints published in 2017 in order to have more consistent data. The geometries contained in the shapefile of the building blueprints allow us to calculate the surface extent of the building rooftops, and its attributes describe the intended use of the buildings, therefore enabling us to remove unsuitable buildings from the dataset. The shape file of the administrative boundaries allowed for us to identify the buildings within each municipality.

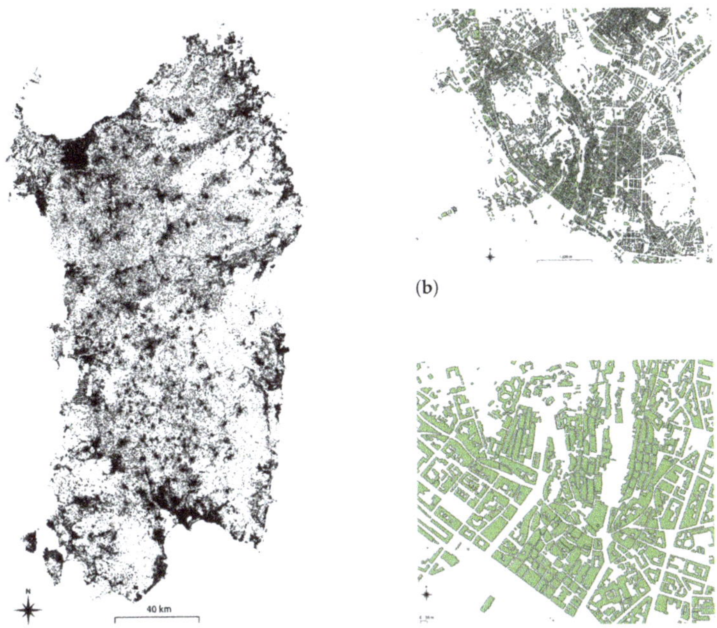

(a) (b) (c)

Figure 1. (**a**) Distribution of buildings (in black) in Sardinia and (**b**,**c**) building blueprints in the Municipality of Cagliari, the administrative center and largest city in Sardinia.

2.3. Elevation Data

We used the elevation data of the buildings, vegetation, and terrain to calculate any projected shadows. In the absence of the three-dimensional model of the buildings, the elevation data is provided by the raster files of the digital surface model (DSM) published by the geoportal of the region of Sardinia [39]. The data covered a good portion of Sardinia (Figure 2a), in particular the coastal, fluvial, and Campidano lowland areas, i.e., the most densely populated territories. The DSMs were obtained from six different LiDAR acquisition campaigns that took place between 2008 and 2013. Belonging to distinct datasets, the raster files are heterogeneous and are offered in a variety of formats (Erdas Imagine, GeoTIFF, ESRI ASCII Grid), each with its own specific peculiarities. As a result, we standardized the raster data on a GeoTIFF format with an EPSG:32632 reference coordinate system and a 1 m horizontal resolution.

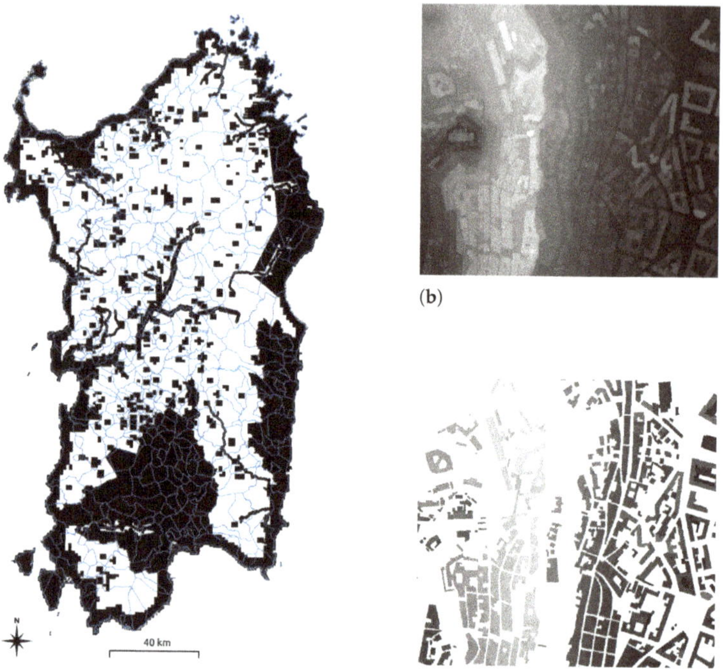

(**a**) (**b**) (**c**)

Figure 2. (**a**) Coverage of the 1 m resolution DSM (in black) over the territory of Sardinia and (**b**,**c**) a clipping of the DSM within the Municipality of Cagliari.

In general, the DSM covers 45.01% of the territory of Sardinia (i.e., 10,844 km^2 out of 24,093 km^2) and as much as 87.36% of the area occupied by buildings (113 km^2 out of 130 km^2), as shown in Figure 3. We assumed that the remaining 12.64% of the buildings had the same characteristics as the buildings for which the digital surface model is available, so we calculate the total results for the individual municipalities and the entire region of Sardinia by means of a proportion.

2.4. Radiation and Weather Data

For the measurements of solar radiation, temperature, and wind speed, we used the data offered by the PVGIS portal [34]. According to the resolution of the PVGIS-SARAH2 solar radiation database [40,41], we identified the 0.05° resolution sectors that overlap the territory of Sardinia (Figure 4) and collected the time series data via a series of application programming interface (API) calls to https://re.jrc.ec.europa.eu/api/v5_2/seriescalc (accessed on 13 November 2022).

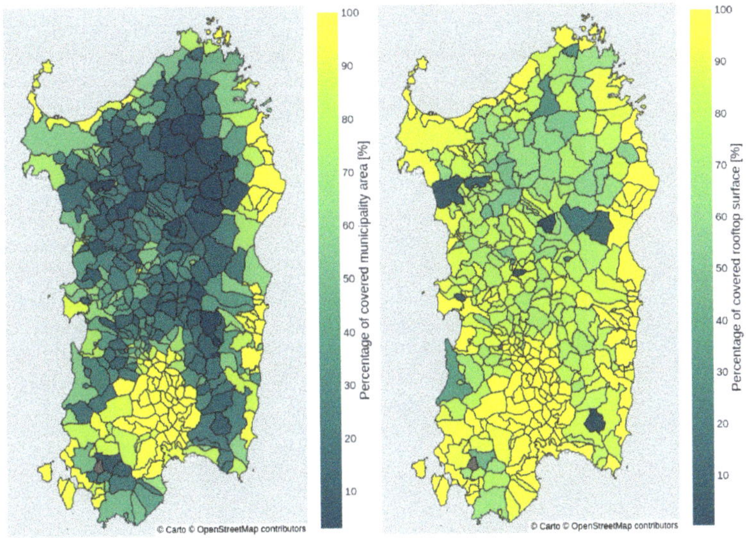

(**a**) (**b**)

Figure 3. (**a**) Percentage of municipality area and (**b**) percentage of rooftop surface covered by the DSM.

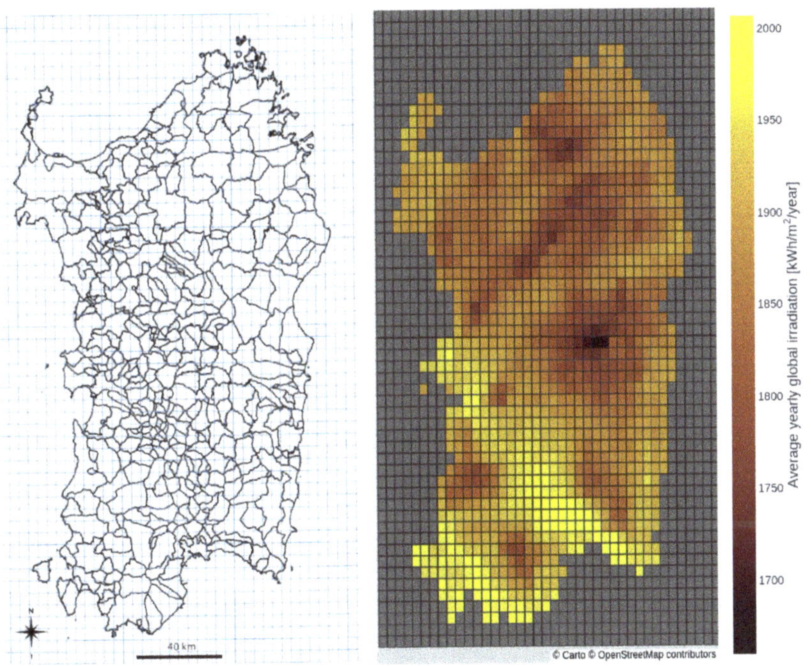

Figure 4. The island of Sardinia represented by the territorial boundaries of its municipalities (in black), (**left**) superimposed on the grid of sectors of size 0.05° (in blue) of the PVGIS-SARAH2 database and (**right**) by the average annual global irradiance on a 1 m^2 surface oriented to the south and inclined by 20°.

Each individual call to the API service returned a 16-year long (2005–2020) hourly time series of the values G_b (DNI_h, direct irradiance on the plane of the array, POA), G_d (DHI_h, diffuse irradiance on the POA), G_r (reflected irradiance on the POA), H_{sun} (elevation of the sun), T_{2m} (air temperature measured at 2 m above the surface) and WS_{10m} (wind speed measured at 10 m above the surface). As a result, each time series contains over 140,000 values.

3. Methodology

In the following paragraphs, we illustrate the methodologies used for computing the slope and aspect of building rooftops (Section 3.1), for evaluating the shadings (Section 3.2), for estimating the radiation incident on rooftops (Section 3.3), for evaluating the energy produced by photovoltaic panels installed on the rooftops themselves (Section 3.4), and for studying the cost-effectiveness of the investment (Section 3.5). An overview of the whole procedure is presented in Figure 5, where the steps within each of the physical, geographic, technical and economic potentials and the connections between different data and methodologies are emphasized.

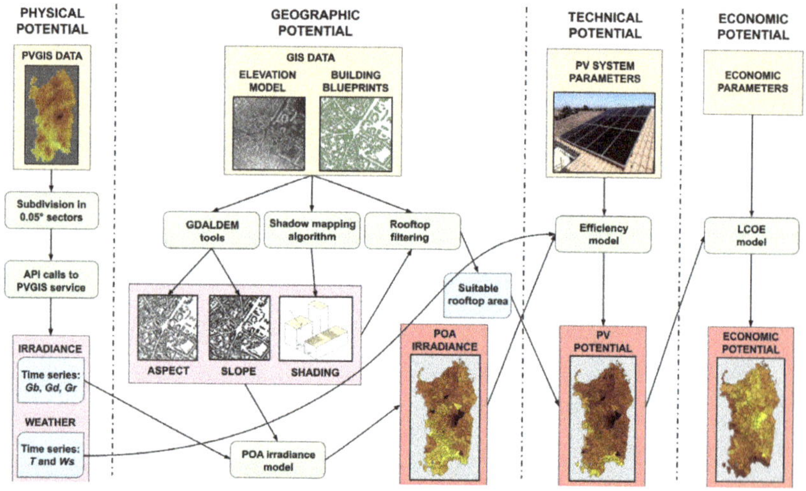

Figure 5. Flowchart representing the workflow followed to obtain the photovoltaic potential of Sardinia. Input data are shown in the yellow boxes, derived data are framed in pink and light-blue boxes, data processing steps were placed in green rounded rectangles, and the workflow results are outlined by red frames.

3.1. Computation of Rooftop Slope and Aspect

By applying the algorithms to the DSM [42], offered by the tool `gdaldem` of the GDAL library [43], we obtained the 1 m resolution raster map estimates of the tilt and orientation angles of the roof surface, denoted respectively with β and γ. These values were necessary for the calculation of the incident irradiance on photovoltaic panels, whose annual total varies significantly depending on the orientation with respect to the south and on the inclination of the rooftops on which they are installed. Figure 6 shows the (a) computed aspect and (b) slope values for the rooftop of the buildings within an area of the Municipality of Cagliari.

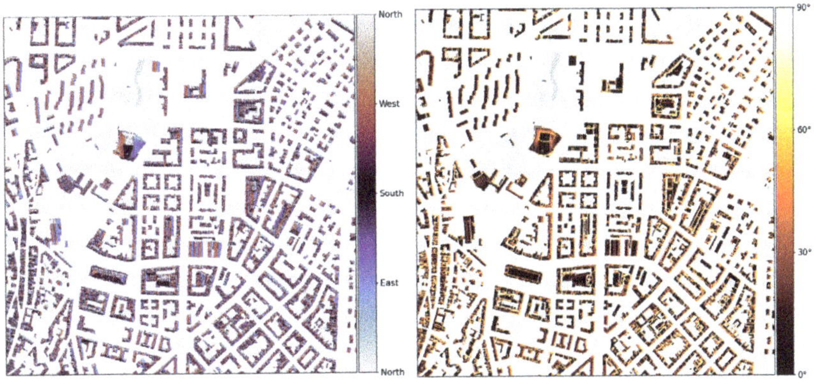

(a) (b)

Figure 6. (**a**) Maps of aspect and (**b**) slope values computed for the building rooftops within the Municipality of Cagliari.

The DSM covers a total building rooftop surface of about 113 km^2, corresponding to 113 million pixels. Not all surfaces are suitable for the installation of panels due to the peculiarities of the buildings, discontinuities, irregularities, and steep inclinations of the rooftop surfaces. After filtering the unsuitable portions of building rooftops, we obtained a total useful surface area of approximately 80 km^2.

Figure 7 shows the cumulative sum of the rooftop surfaces: about 60% of building rooftops were characterized by a slope less than 20°. Around 25% of rooftops had a slope greater than 45° and were thus excluded from the dataset. A further 5% of the rooftops were removed because they were part of contiguous surfaces that were too small to allow for the installation of photovoltaic modules. The aspect of the building rooftops is instead described by a continuous uniform distribution (not shown).

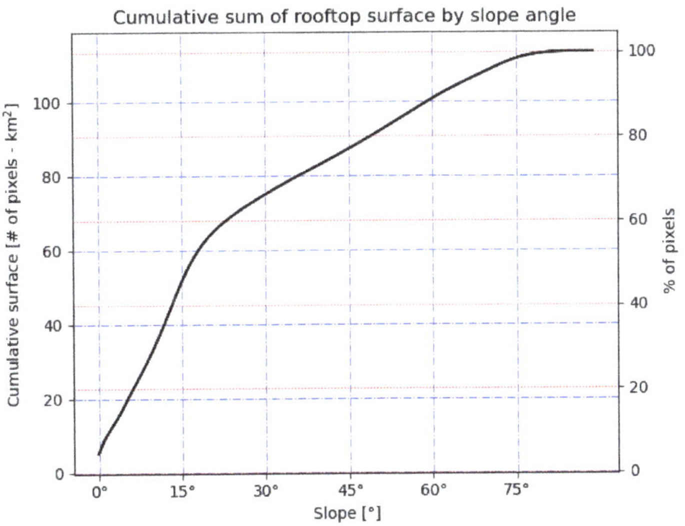

Figure 7. Cumulative sum of rooftop surface, sorted by increasing tilt angle.

3.2. Computation of Shadows over Rooftops

Knowing whether a portion of a rooftop is shaded for a given position of the sun in the sky allows the amount of total irradiation received by the photovoltaic panels to be corrected by excluding the contribution of the direct component of the irradiation. Using the shadow mapping algorithm presented in [32], we determined for all individual pixels of the rooftop surfaces whether they are illuminated or in shadow for each of the selected positions of the sun. The procedure is a variation of the well-known Bresenham algorithm [44] used in computer graphics to draw segments between two points on a grid. Please refer to [32] for a detailed description of our implementation.

3.3. Estimation of Incident Radiation

The mean annual irradiance on the plane of the array E^y_{POA} was obtained from the sum of several contributions, i.e., the mean annual irradiances due to the direct component E^y_b, the diffuse component from the sky E^y_d and the diffuse component from the ground E^y_g:

$$E^y_{\text{POA}} = E^y_b + E^y_d + E^y_g \tag{1}$$

The individual addends of Equation (1) could be approximated to summations:

$$E^y_b = \int_0^T E_b(t)dt \approx \Delta t \sum_{i=1}^{N} E_{b,i} = \Delta t \sum_{i=1}^{N} F_i \text{DNI}_i \cos(\text{AOI}_i) \tag{2}$$

Equation (2) calculates the average annual contribution E^y_b due to direct irradiance $E_b(t)$ over a year (T), which was discretized into one-hour time intervals Δt. In Equation (2) DNI_i is the mean direct normal irradiance, $F_i = \{0, 1\}$ indicates whether the point at time instant t_i of the year is in shadow or is illuminated by the sun, and AOI_i is the mean angle of incidence of the solar ray with respect to the panel surface at time t_i.

$$E^y_d = \int_0^T E_d(t)dt \approx \Delta t \sum_{i=1}^{N} E_{d,i} = \Delta t \frac{1 + \cos\beta}{2} \sum_{i=1}^{N} \text{DHI}_i \tag{3}$$

Equation (3) is relative to the average annual contribution E^y_d due to diffuse irradiance $E_d(t)$ from the sky on a surface tilted with angle β with the isotropic sky model [45]; DHI_i is the diffuse horizontal irradiance.

$$E^y_g = \int_0^T E_g(t)dt \approx \Delta t \sum_{i=1}^{N} E_{g,i} = \Delta t \frac{1 - \cos\beta}{2} \sum_{i=1}^{N} \rho \text{GHI}_i \tag{4}$$

Equation (4) is relative to the average annual contribution E^y_g due to diffuse irradiance from the ground $E_g(t)$ on a tilted surface; GHI_i is the global horizontal irradiance, and ρ is the albedo.

In our model, the values of GHI, DHI, and DNI depend solely on the position of the Sun with respect to the panel. Each instant t_i of the year corresponds to a specific position of the sun in the sky, as defined by azimuth angles α_i and zenith angles θ_i. Performing the calculation described in Section 3.2 for millions of pixels and for hundreds of thousands of possible positions of the Sun in the sky (a year consists of over 500,000 min) is too computationally onerous. With the objective of keeping the spatial resolution of 1 m unchanged, it is possible to significantly reduce the number of sun positions for which to calculate whether a point is illuminated by the sun or is in shadow.

The employed strategy was to divide the sky into sectors with a resolution of $\Delta\alpha = \Delta\theta = 5°$ (both in azimuth and zenith) without sacrificing accuracy [32]. As a result, Equations (2)–(4) are transformed into:

$$E_b^y = \Delta t \sum_{i=1}^{N} F_i \mathrm{DNI}_i \cos(\mathrm{AOI}_i) = \sum_{p=1,q=1}^{N_p,N_q} F_{p,q} E_{p,q}^{\mathrm{DNI}} \cos(\mathrm{AOI}_{p,q}) \tag{5}$$

$$E_d^y = \Delta t \frac{1+\cos\beta}{2} \sum_{i=1}^{N} \mathrm{DHI}_i = \frac{1+\cos\beta}{2} \sum_{p=1,q=1}^{N_p,N_q} E_{p,q}^{\mathrm{DHI}} \tag{6}$$

$$E_g^y = \Delta t \rho \frac{1-\cos\beta}{2} \sum_{i=1}^{N} \mathrm{GHI}_i = \rho \frac{1-\cos\beta}{2} \sum_{p=1,q=1}^{N_p,N_q} E_{p,q}^{\mathrm{GHI}} \tag{7}$$

where $N_p, N_q \ll N$, and, using Iverson's notation,

$$E_{p,q}^{\mathrm{DNI}} = \Delta t \sum_{\alpha_i,\theta_i} [\Delta\alpha(p-1) \le \alpha_i < \Delta\alpha p][\Delta\theta(q-1) \le \theta_i < \Delta\theta q]\mathrm{DNI}_i \tag{8}$$

$$E_{p,q}^{\mathrm{DHI}} = \Delta t \sum_{\alpha_i,\theta_i} [\Delta\alpha(p-1) \le \alpha_i < \Delta\alpha p][\Delta\theta(q-1) \le \theta_i < \Delta\theta q]\mathrm{DHI}_i \tag{9}$$

$$E_{p,q}^{\mathrm{GHI}} = \Delta t \sum_{\alpha_i,\theta_i} [\Delta\alpha(p-1) \le \alpha_i < \Delta\alpha p][\Delta\theta(q-1) \le \theta_i < \Delta\theta q]\mathrm{GHI}_i \tag{10}$$

Equations (8)–(10) allow for the calculation of the average direct, diffuse, and global components of solar irradiance for each portion of the sky as a function of the p, q indices of azimuth and zenith.

As a result, the shading calculation for each pixel only needs to be performed a few hundred times, i.e., for combinations of α_p and θ_q angles for which $\mathrm{DNI}_{p,q} > 0$. By repeating the computations for each pixel of the DSM raster belonging to the area inside the perimeter of the buildings, we thus obtain the 1 m resolution map of the mean annual irradiance on the plane of the array E_{POA}^y, and the components E_b^y, E_d^y and E_g^y for all the buildings in Sardinia.

Figure 8a shows an extract of the global horizontal irradiance (GHI) map, whereas Figure 8b depicts the irradiance on the plane of the array (POA). It is immediately noticeable how the darker areas, i.e., those facing north or shaded by other buildings, receive less radiation than the light-colored areas do, i.e., those facing south or are taller than the surrounding buildings.

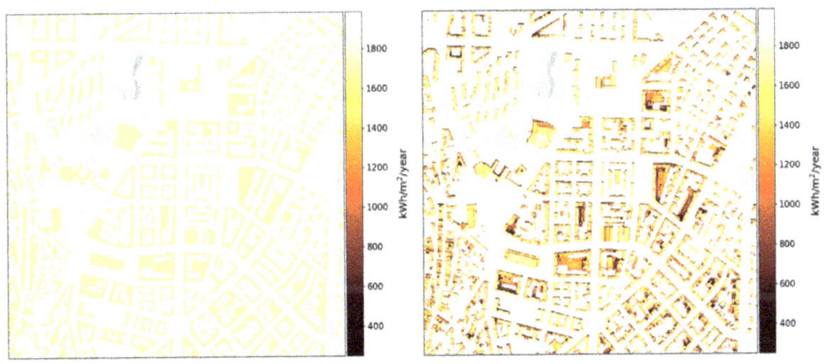

(a) (b)

Figure 8. *Cont.*

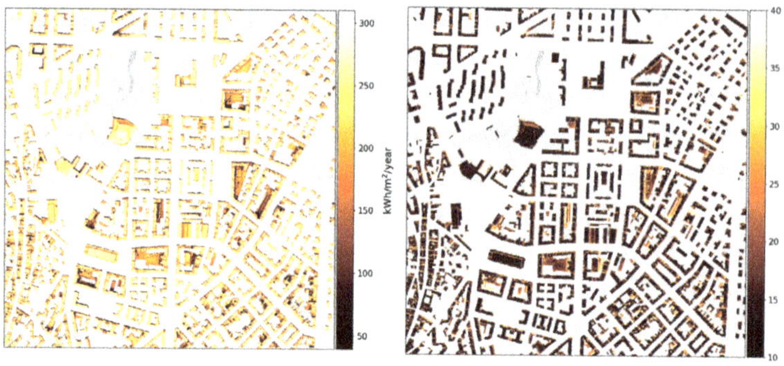

(**c**) (**d**)

Figure 8. Maps showing, for the rooftop surfaces within the Municipality of Cagliari, (**a**) the yearly average values per m^2 of GHI, (**b**) irradiation on the POA, (**c**) and photovoltaic potential; (**d**) average LCOE.

3.4. Estimation of Photovoltaic Technical Potential

For the estimation of the photovoltaic technical potential, historical time series of temperature and wind speed provided by PVGIS were used in addition to the irradiation on the POA. The power output of the photovoltaic installations, according to the model of Huld et al. [35], depends on irradiation, weather conditions, system characteristics, and system losses:

$$P_{\text{pv}} = P_{\text{POA}} \cdot P_{\text{STC}} \cdot \eta_r \cdot (1 - L) \tag{11}$$

where P_{POA} represents the irradiance in the plane of the arrays, and P_{STC} represents the power output from the photovoltaic modules determined under standard test conditions; in our case, it defined as $P_{\text{STC}} = A \cdot \eta_n = 182$ W, where $\eta_n = 0.182$ represents the nominal (18.2%) efficiency [33] and $A = 1$ m^2 is the area covered by a single pixel of the raster. Relative efficiency η_r is calculated as follows:

$$\eta_r = 1 + k_1 \log(P') + k_2 \log(P')^2 + k_3 T_m + k_4 T_m \log(P') + k_5 T_m \log(P')^2 + k_6 T_m^2 \tag{12}$$

where $P' = P_{\text{POA}}/1000$ and parameters k_i, which depend on the solar panel technology used, were calibrated on the basis of measurements by ESTI, the Joint Research Centre's European Solar Test Installation laboratory [35]. Temperature module T_m is defined as follows:

$$T_m = T_a - 25^\circ + \frac{P_{\text{POA}}}{U_0 + U_1 \cdot W_s} \tag{13}$$

where T_a is the ambient temperature [46], W_s is the wind speed, and coefficients $U_0 = 26.9$ and $U_1 = 6.2$ are determined in [47]. Lastly, generic system loss $L = 14.0\%$, as recommended by PVGIS, was considered. Such system losses include the reduction of the panel efficiency over time and the losses due to the transfer of power from the photovoltaic panels to the electrical grid via cables and inverter [7].

By integrating power output P_{pv} from Equation (11) over a one-year period, we obtained the average energy E_{pv} generated annually by a hypothetical 1 m^2 sized panel installed on a portion of a roof characterized by specific irradiation values, weather conditions, tilt, and orientation. Thus, we obtained the technical potential map, an extract of which is shown as an example in Figure 8c, where each pixel represents the amount of energy E_{pv} that could be produced annually by photovoltaic modules characterized by a 1 m^2 surface, efficiency of 18.2% and system losses of 14.0%.

3.5. Estimation of Economic Potential

An initial estimation of the economic potential is obtained by calculating the levelized cost of electricity (LCOE), which expresses the ratio of the costs incurred for the installation of a photovoltaic system to the energy produced by the system itself over the period of operation. Therefore, it indicates how much it would cost to produce 1 kWh with a 1 kWp photovoltaic system. When the LCOE value is lower than the cost of a kWh purchased from the grid, the investment is remunerative. We calculated the LCOE coefficient as follows [36,48]:

$$\mathrm{LCOE} = \frac{\mathrm{CAPEX} + \sum_{n=1}^{N} (\mathrm{OPEX}_n - \mathrm{TD}_n)(1+r)^{-n}}{\sum_{n=1}^{N} E_{\mathrm{kWp},n}^{\mathrm{y}} (1+r)^{-n}} \tag{14}$$

$$E_{\mathrm{kWp}}^{\mathrm{y}} = E_{\mathrm{pv}} \cdot A/\eta_n = 5.5 \cdot E_{\mathrm{pv}} \tag{15}$$

where CAPEX $= 2000$ €/kWp (capital expenditure) is the initial cost of the plant [49], which includes the purchase of panels and inverters, and the installation costs; OPEX $= 50$ €/kWp (operating expense) is the annual running cost of the system [49] that takes into account the tariff to be paid to the grid operator and the eventual cleaning and maintenance of the system; TD represents the tax deductions or purchase incentives that for simplicity we consider null; The cost of capital is $r = 0.05$ [49]; we assumed an operating life of $N = 20$ years. Lastly, $E_{\mathrm{kWp}}^{\mathrm{y}}$ represents the average energy produced annually by a 1 kWp photovoltaic system with $\eta_n = 18.2\%$ [33].

It is possible to map the cost of electricity by calculating the LCOE for all the pixels on building rooftops, as shown in Figure 8d for a district of the Municipality of Cagliari. Dark pixels show the rooftop surfaces where the installation of photovoltaic systems is most economically convenient. The choice of colours renders the map chromatically complementary to the technical potential map shown in Figure 8c.

4. Results

Using the methodologies presented in Section 3 we computed, with a spatial resolution of 1 m and for the whole territory of Sardinia, the shape (slope and aspect) of the building rooftops, the irradiation incident on the rooftops (taking into account any shading due to the presence of nearby buildings or other obstructions), the average annual energy delivered by standard photovoltaic panels hypothetically installed on the rooftops, and the cost to be incurred for installing a 1 kWp photovoltaic system in order for it to produce 1 kWh of energy. The results provide an overview of the rooftop photovoltaic potential in Sardinia.

We estimated a yearly production of 22 TWh of electricity from photovoltaic systems for the entire island of Sardinia. Its two most populous cities, Cagliari and Sassari, contribute almost 1 TWh each. The other major cities of Olbia, Quartu Sant'Elena, Alghero, Oristano, and Nuoro, each exceeded a yearly production of 300 GWh (Table 1 and Figure 9a).

Table 1. Summary table of annual photovoltaic potential by number of municipalities.

PV Potential (GWh)	Number of Municipalities
$E_{\mathrm{pv}} \leq 50$	257
$50 < E_{\mathrm{pv}} \leq 100$	68
$100 < E_{\mathrm{pv}} \leq 200$	35
$200 < E_{\mathrm{pv}} \leq 400$	11
$400 < E_{\mathrm{pv}} \leq 800$	4
$800 < E_{\mathrm{pv}}$	2

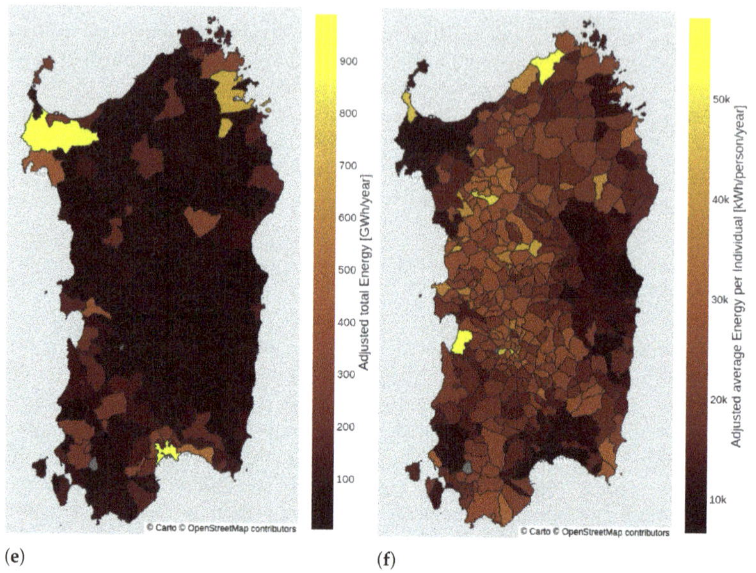

Figure 9. (**a**) Total yearly photovoltaic potential and (**b**) average yearly photovoltaic potential per resident for the municipalities in Sardinia.

Figure 9b shows how much energy could be produced per resident per year after the installation of photovoltaic modules on suitable rooftops. Depending on the municipality in Sardinia, the technical potential varied from 215 to 265 kWh/m^2/year, and from 10 to 60 MWh/individual/year. The per capita potential depends on two main factors: the presence of apartment buildings, which are widespread in larger cities, lowering the amount of rooftop area per individual; the presence of vacation houses in tourist areas, which are unused for most of the year, and whose owners are rarely among the residents, increasing the available surface per capita.

Compared to the total value, which is strongly related to the extension of the municipality's territory and the number of residents, the photovoltaic potential expressed per m^2 is more indicative of the potential of each individual area. Figure 10a shows the annual average global horizontal irradiance per m^2, as estimated by PVGIS. The map is similar to Figure 4b, but instead of the 0.05° sectors determined by the PVGIS-SARAH2 database, the GHI values are distributed over the individual territories (from 1600 to close to 1800 kWh/m^2/year) The reduction in irradiance on the rooftops due to the shading, inclination, and orientation of the panels leads to the map in Figure 10b, which shows the actual irradiance received per m^2 on the plane of the arrays. For some municipalities, the change is drastic, possibly justified by the fortunate exposure of most buildings to the sun. The towns that receive lower irradiance are either in mountainous locations (for which the average irradiance is lower than that for locations on the plains) or are located in deep valleys, situations that greatly reduce the irradiance received when the Sun is not yet high in the sky.

The map of the photovoltaic potential in Figure 10c is qualitatively very similar to the previous one; in fact, the (small) differences between one municipality and another are explained by the equally small differences in temperature and wind speed. Lastly, Figure 10d depicts the average levelized cost of electricity in Sardinia for all municipalities ranging from 14 to 21 EuroCent per kWh. Should the cost of energy remain high as it has been, installing photovoltaic systems in the vast majority of the buildings might be worthwhile, and the energy surplus can be sold to other regions.

The estimated annual photovoltaic potential for the island of Sardinia as a whole, equal to 22 TWh, is obtained by computing the energy production from photovoltaic modules distributed over a surface area larger than 92 km^2 (on a total rooftop area of almost 130 km^2). This equates to an average annual production of 238.92 kWh per m^2, or 1314 kWh produced annually for each kWp installed, using the technology chosen as an example.

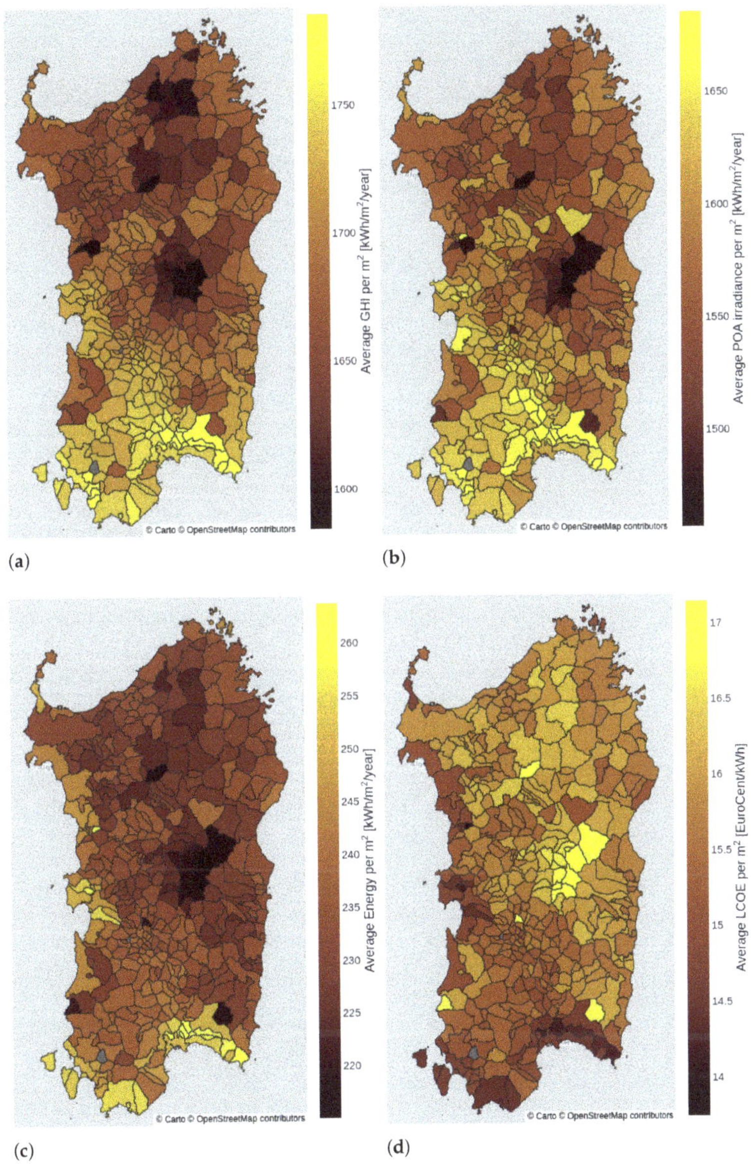

Figure 10. (**a**) Average yearly GHI per m^2, (**b**) irradiation on the POA per m^2, (**c**) photovoltaic potential per m^2 and (**d**) average LCOE for the municipalities in Sardinia.

Figure 11 shows the dependence of annual photovoltaic potential, per m^2, on rooftop orientation and tilt. In Sardinia, the optimal rooftops which guarantee about 250 $kWh/m^2/year$ are those with an orientation between south-west and south-east and a slope between 10° and

30°. North-facing façades are unproductive unless they have an inclination of less than 20°, in which case a production of approximately 200 kWh/m^2/year is expected.

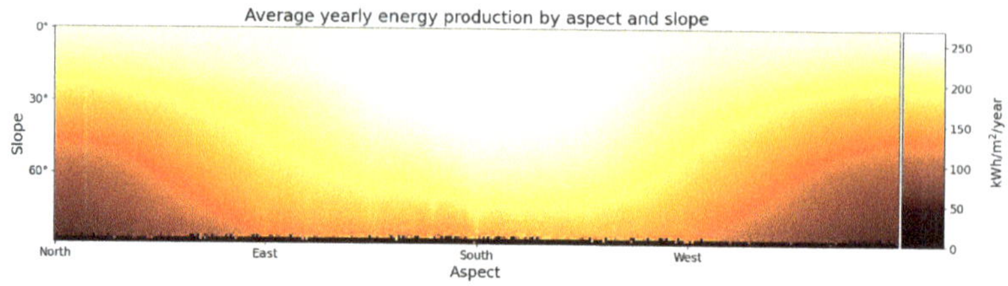

Figure 11. Average annual energy production per m^2 based on the orientation and inclination of photovoltaic panels in Sardinia.

5. Discussion

Our procedure estimated a yearly production of 239 kWh per m^2, i.e., a photovoltaic potential of 22 TWh distributed over a rooftop area of about 92 km^2. The JRC [7] estimates, for the region of Sardinia, a production of 4.68 TWh, but over an area of just 31.96 km^2, corresponding to an average of 146.41 kWh per m^2. It is not explicit in the text, but it would appear that they used a panel efficiency of 10% instead of 18.2% (as in our case). Using the 18.2% efficiency (5.5 m^2/kWp instead of 10), results in an annual production of 266.19 kWh/m^2. The JRC estimate was, thus, higher than ours, although this can be explained by their characterization of the modules with optimal aspect (due south) and slope (20°) and not, as in our case, with the actual tilt and orientation of the rooftops. The JRC estimate, moreover, is based on much coarser analysis that does not take into account the actual location of the buildings or the orientation of the rooftops.

Sardinia's electricity production in 2019 (pre-COVID-19) was 13.14 TWh, with photovoltaic systems accounting for only 8.8% (1.15 TWh), while total consumption was 8.47 TWh [50]. As for the domestic and tertiary sectors, consumption in 2019 amounted to 4.43 TWh. The photovoltaic potential of Sardinia is therefore capable of easily covering domestic and tertiary sector consumption, and that in the industrial and agricultural sectors. Consumption and potential per province are detailed in Table 2. The smaller available surface area per inhabitant and the higher concentration of tertiary sector enterprises render the province of the metropolitan city of Cagliari the most energy-intensive with respect to its photovoltaic potential. This is despite the fact that the installation of modules on just one third of the total rooftop surface could meet the energy needs of all households and services.

Table 2. Electricity consumption in the domestic and services sectors, and photovoltaic potential for the 5 provinces of Sardinia. The percentage indicates the fraction of eligible rooftops that, if equipped with a photovoltaic system and accumulators, would cover the needs of the domestic and services sectors.

Province	Energy Demand	PV Potential	% of Rooftop Surface
Cagliari	1248 GWh	3.80 TWh	32.84%
Nuoro	482 GWh	2.95 TWh	16.34%
Oristano	374 GWh	3.15 TWh	11.87%
Sassari	1514 GWh	6.65 TWh	22.77%
Sud Sardegna	810 GWh	5.49 TWh	14.75%
Sardinia	4.43 TWh	22.04 TWh	20.10%

The distribution of the estimated values for the levelized cost of electricity is shown in Figure 12, where 50% of installations had an LCOE below 14.76 EuroCent/kWh, and 50% of installations had an LCOE between 13.94 and 16.10 EuroCent/kWh.

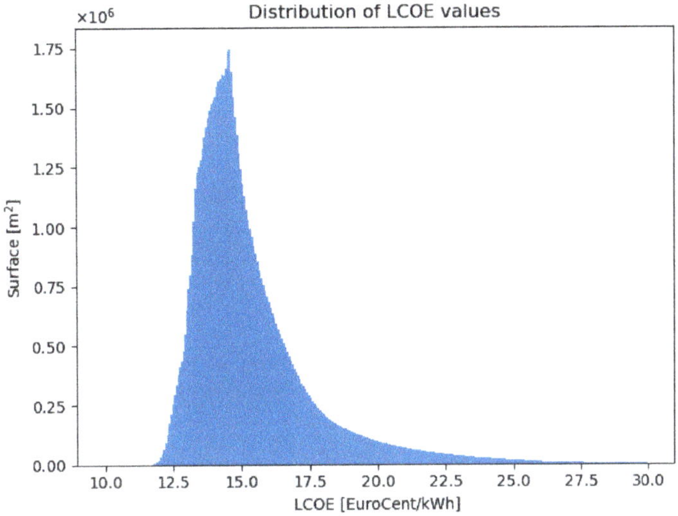

Figure 12. Distribution of the levelized cost of electricity for all the rooftop surfaces in Sardinia.

Figure 13 shows the total energy from rooftop PV as a function of the corresponding levelized cost of electricity, and shows that almost all of the available roof area could be exploited for an LCOE of 17.5 EuroCent/kWh.

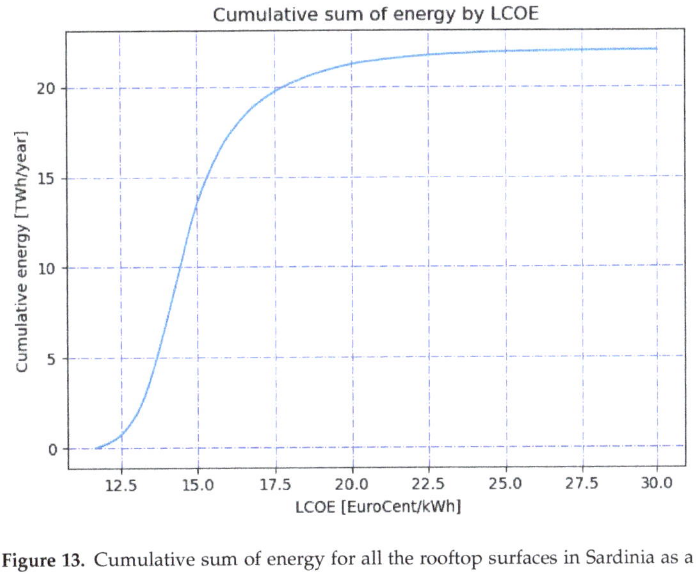

Figure 13. Cumulative sum of energy for all the rooftop surfaces in Sardinia as a function of the levelized cost of electricity.

In Italy, electricity prices in 2022 soared above 30 EuroCent per kWh, making photovoltaic system installation worthwhile even in the least productive locations [51]. Assuming a return to the situation prior to COVID and to quieter geopolitical circumstances, the average energy price from 2010 to 2019 was 9.45 EuroCent/kWh, well below our LCOE values. However, the LCOE values were calculated without taking into account the incentives (SuperBonus 110% or 50% tax deduction, preferential VAT, on-site exchange, -i.e., the sale of energy) thanks to which the installation would be worthwhile even with a return of energy prices to pre-2020 values.

Areas for the improvement of our estimate mainly concern the quality of the data available for analysis. The data used in the work were not particularly recent: the building blueprints, despite having been digitally published in 2017, were obtained from regional databases compiled between 1994 and 2000, and partially updated in 2006, while the digital surface model was based on LiDAR observations performed between 2008 and 2013. Therefore, there is a discrepancy between the data shown by the raster and the cadastral data (e.g., the raster of 2013 might show a building not yet registered in the blueprint dataset); the difference between the current situation (2022) and the one described by the data might not be negligible (entire neighbourhoods might be built in a decade). In terms of data accuracy, the digital surface model does not account for the presence of empty spaces; sometimes the building blueprints also include internal courtyards within the perimeter (e.g., see Figure 14); situations such as these are usually discarded due to the presence of vegetation that makes the presumed building coverage irregular. The available data do not yet cover the entire territory of Sardinia or even the entire set of buildings; the overall estimate was obtained by rescaling the calculation made on the covered area, but it is evident that a complete coverage would lead to fewer uncertainties.

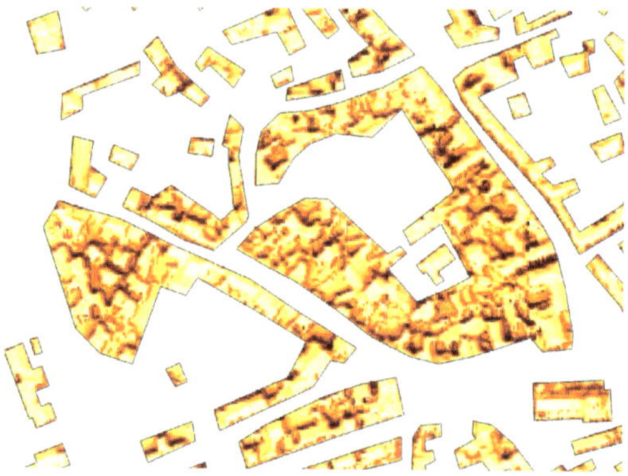

Figure 14. Portion of the irradiation map in which the perimeter of some buildings does not follow the actual perimeter of the buildings themselves, but also erroneously includes the inner courtyards, thus misleading the algorithm that considers vegetation, for example, as an irregular cover. Surfaces with a slope exceeding 45° and not belonging to contiguous areas of at least 30 m^2 are, in any case, not considered in the calculation of municipal and regional photovoltaic potential.

6. Conclusions

In this paper, we presented a high-resolution rooftop photovoltaic potential estimation procedure based on GIS and satellite irradiance data, highly optimized to allow for calculation over large areas.

We applied the procedure to the case of the Sardinia region in Italy. This gave us an estimate of the geographic, technical, and economic potential for the entire region with a spatial resolution of 1 m.

The computation provided an estimate of photovoltaic potential on building roofs as a function of the cost of electricity, with an upper limit value of 22 TWh per year. This potential would abundantly cover the energy needs of the region, in particular those related to household consumption and for service activities directly associated with the buildings on which the panels could be installed. Energy is not power, however, and the need to instantly balance demand and generation of electric power by ensuring the balance of the distribution system is clearly left out of this analysis. This work focused on estimating geographic potential as a basis for further processing, and the technical and economic analysis is based on standard technologies. We, therefore, make our results available to the community in the hope that other researchers could further investigate aspects of economic potential in relation to the different PV technologies available, and in relation to a changing regulatory and economic framework. The procedure is replicable in other areas where the necessary data are available. Our research will continue with the development of artificial intelligence techniques for identifying suitable areas for panel installation from satellite images.

Author Contributions: Conceptualization, A.P. and L.M.; methodology, A.P. and L.M.; software, A.P.; validation, A.P. and L.M.; formal analysis, L.M.; investigation, A.P.; resources, A.P.; data curation, A.P.; writing—original draft preparation, A.P.; writing—review and editing, L.M.; visualization, A.P.; supervision, L.M.; project administration, L.M.; funding acquisition, L.M. All authors have read and agreed to the published version of the manuscript.

Funding: This research was funded by the Tessuto Digitale Metropolitano (TDM) project with funds provided by the Autonomous Region of Sardinia through Action 1.2.2 of the POR FESR Sardinia 2014–2020.

Institutional Review Board Statement: Not applicable.

Informed Consent Statement: Not applicable.

Data Availability Statement: The results presented in this paper are available upon request and are shown at http://s.crs4.it/JM (accessed on 13 November 2022).

Conflicts of Interest: The authors declare no conflict of interest.

Abbreviations

The following abbreviations are used in this manuscript:

AOI	Angle of incidence
API	Application programming interface
CAPEX	Capital expenditure
DBGT	Geotopographic database
DHI	Diffuse horizontal irradiance
DNI	Direct normal irradiance
DSM	Digital surface model
GHI	Global horizontal irradiance
GIS	Geographic information system
LCOE	Levelized cost of electricity
LiDAR	Light detection and ranging
OPEX	Operating expense
POA	Plane of the array
PV	Photovoltaic
STC	Standard test conditions
TD	Tax deductions

References

1. European Commission. Directive 2009/28/EC of the European Parliament and of the Council of 23 April 2009 on the promotion of the use of energy from renewable sources and amending and subsequently repealing Directives 2001/77/EC and 2003/30/EC. *Off. J. Eur. Union* **2009**, *5*, 2009.
2. European Commission. Directive (EU) 2018/2001 of the European Parliament and of the Council of 11 December 2018 on the promotion of the use of energy from renewable sources. *Off. J. Eur. Union* **2018**, *L328*, 82.
3. European Commission. Communication from the Commission to the European Parliament, the European Council, the European Economic and Social Committee and the Committee of the Regions—REPowerEU Plan. Available online: https://eur-lex.europa.eu/legal-content/EN/TXT/?uri=COM%3A2022%3A230%3AFIN (accessed on 13 November 2022).

4. Italian Regulatory Authority for Energy, Networks and Environment (ARERA). Produzione Lorda di Energia Elettrica per Fonte. Available online: https://www.arera.it/it/dati/eem6.htm# (accessed on 13 November 2022).
5. Ascione, F.; Bianco, N.; Mauro, G.M.; Napolitano, D.F.; Vanoli, G.P. Comprehensive analysis to drive the energy retrofit of a neighborhood by optimizing the solar energy exploitation–An Italian case study. *J. Clean. Prod.* **2021**, *314*, 127998. [CrossRef]
6. Calise, F.; Cappiello, F.L.; Cimmino, L.; d'Accadia, M.D.; Vicidomini, M. Dynamic modelling and thermoeconomic analysis for the energy refurbishment of the Italian building sector: Case study for the "Superbonus 110%" funding strategy. *Appl. Therm. Eng.* **2022**, *213*, 118689. [CrossRef]
7. Bódis, K.; Kougias, I.; Jäger-Waldau, A.; Taylor, N.; Szabó, S. A high-resolution geospatial assessment of the rooftop solar photovoltaic potential in the European Union. *Renew. Sustain. Energy Rev.* **2019**, *114*, 109309. [CrossRef]
8. Jäger-Waldau, A. The untapped area potential for photovoltaic power in the European Union. *Clean Technol.* **2020**, *2*, 440–446. [CrossRef]
9. Hoang, A.T.; Nguyen, X.P. Integrating renewable sources into energy system for smart city as a sagacious strategy towards clean and sustainable process. *J. Clean. Prod.* **2021**, *305*, 127161. [CrossRef]
10. Backe, S.; Zwickl-Bernhard, S.; Schwabeneder, D.; Auer, H.; Korpås, M.; Tomasgard, A. Impact of energy communities on the European electricity and heating system decarbonization pathway: Comparing local and global flexibility responses. *Appl. Energy* **2022**, *323*, 119470. [CrossRef]
11. Wierling, A.; Zeiss, J.P.; Lupi, V.; Candelise, C.; Sciullo, A.; Schwanitz, V.J. The contribution of energy communities to the upscaling of photovoltaics in Germany and Italy. *Energies* **2021**, *14*, 2258. [CrossRef]
12. Komninos, N. Net Zero Energy Districts: Connected Intelligence for Carbon-Neutral Cities. *Land* **2022**, *11*, 210. [CrossRef]
13. Fakhraian, E.; Forment, M.A.; Dalmau, F.V.; Nameni, A.; Guerrero, M.J.C. Determination of the urban rooftop photovoltaic potential: A state of the art. *Energy Rep.* **2021**, *7*, 176–185. [CrossRef]
14. Fakhraian, E.; Alier, M.; Valls Dalmau, F.; Nameni, A.; Casañ Guerrero, M.J. The Urban Rooftop Photovoltaic Potential Determination. *Sustainability* **2021**, *13*, 7447. [CrossRef]
15. Castellanos, S.; Sunter, D.A.; Kammen, D.M. Rooftop solar photovoltaic potential in cities: How scalable are assessment approaches? *Environ. Res. Lett.* **2017**, *12*, 125005. [CrossRef]
16. Joshi, S.; Mittal, S.; Holloway, P.; Shukla, P.R.; Ó Gallachóir, B.; Glynn, J. High resolution global spatiotemporal assessment of rooftop solar photovoltaics potential for renewable electricity generation. *Nat. Commun.* **2021**, *12*, 5738. [CrossRef]
17. Hołuj, A.; Ilba, M.; Lityński, P.; Majewski, K.; Semczuk, M.; Serafin, P. Photovoltaic Solar Energy from Urban Sprawl: Potential for Poland. *Energies* **2021**, *14*, 8576. [CrossRef]
18. Bernasconi, D.; Guariso, G. Rooftop PV: Potential and Impacts in a Complex Territory. *Energies* **2021**, *14*, 3687. [CrossRef]
19. Walch, A.; Castello, R.; Mohajeri, N.; Scartezzini, J.L. Big data mining for the estimation of hourly rooftop photovoltaic potential and its uncertainty. *Appl. Energy* **2020**, *262*, 114404. [CrossRef]
20. Todeschi, V.; Mutani, G.; Baima, L.; Nigra, M.; Robiglio, M. Smart solutions for sustainable cities—The re-coding experience for harnessing the potential of urban rooftops. *Appl. Sci.* **2020**, *10*, 7112. [CrossRef]
21. Borfecchia, F.; Caiaffa, E.; Pollino, M.; De Cecco, L.; Martini, S.; La Porta, L.; Marucci, A. Remote Sensing and GIS in planning photovoltaic potential of urban areas. *Eur. J. Remote Sens.* **2014**, *47*, 195–216. [CrossRef]
22. Aslani, M.; Seipel, S. Automatic identification of utilizable rooftop areas in digital surface models for photovoltaics potential assessment. *Appl. Energy* **2022**, *306*, 118033. [CrossRef]
23. Nguyen, H.T.; Pearce, J.M. Incorporating shading losses in solar photovoltaic potential assessment at the municipal scale. *Sol. Energy* **2012**, *86*, 1245–1260. [CrossRef]
24. Montealegre, A.; García-Pérez, S.; Guillén-Lambea, S.; Monzón-Chavarrías, M.; Sierra-Pérez, J. GIS-based assessment for the potential of implementation of food-energy-water systems on building rooftops at the urban level. *Sci. Total Environ.* **2022**, *803*, 149963. [CrossRef] [PubMed]
25. Koch, H.; Lechner, S.; Erdmann, S.; Hofmann, M. Assessing the Potential of Rooftop Photovoltaics by Processing High-Resolution Irradiation Data, as Applied to Giessen, Germany. *Energies* **2022**, *15*, 6991. [CrossRef]
26. de Vries, T.N.; Bronkhorst, J.; Vermeer, M.; Donker, J.C.; Briels, S.A.; Ziar, H.; Zeman, M.; Isabella, O. A quick-scan method to assess photovoltaic rooftop potential based on aerial imagery and LiDAR. *Sol. Energy* **2020**, *209*, 96–107. [CrossRef]
27. Krapf, S.; Kemmerzell, N.; Khawaja Haseeb Uddin, S.; Hack Vázquez, M.; Netzler, F.; Lienkamp, M. Towards scalable economic photovoltaic potential analysis using aerial images and deep learning. *Energies* **2021**, *14*, 3800. [CrossRef]
28. Fuster-Palop, E.; Prades-Gil, C.; Masip, X.; Viana-Fons, J.D.; Payá, J. Innovative regression-based methodology to assess the techno-economic performance of photovoltaic installations in urban areas. *Renew. Sustain. Energy Rev.* **2021**, *149*, 111357. [CrossRef]
29. Sun, T.; Shan, M.; Rong, X.; Yang, X. Estimating the spatial distribution of solar photovoltaic power generation potential on different types of rural rooftops using a deep learning network applied to satellite images. *Appl. Energy* **2022**, *315*, 119025. [CrossRef]
30. Robinson, D. Urban morphology and indicators of radiation availability. *Sol. Energy* **2006**, *80*, 1643–1648. [CrossRef]
31. Rodríguez, L.R.; Duminil, E.; Ramos, J.S.; Eicker, U. Assessment of the photovoltaic potential at urban level based on 3D city models: A case study and new methodological approach. *Sol. Energy* **2017**, *146*, 264–275. [CrossRef]

32. Pinna, A.; Massidda, L. A procedure for complete census estimation of rooftop photovoltaic potential in urban areas. *Smart Cities* **2020**, *3*, 873–893. [CrossRef]
33. Bódis, K.; Kougias, I.; Taylor, N.; Jäger-Waldau, A. Solar photovoltaic electricity generation: A lifeline for the European coal regions in transition. *Sustainability* **2019**, *11*, 3703. [CrossRef]
34. Huld, T.; Müller, R.; Gambardella, A. A new solar radiation database for estimating PV performance in Europe and Africa. *Sol. Energy* **2012**, *86*, 1803–1815. [CrossRef]
35. Huld, T.; Friesen, G.; Skoczek, A.; Kenny, R.P.; Sample, T.; Field, M.; Dunlop, E.D. A power-rating model for crystalline silicon PV modules. *Sol. Energy Mater. Sol. Cells* **2011**, *95*, 3359–3369. [CrossRef]
36. Branker, K.; Pathak, M.; Pearce, J.M. A review of solar photovoltaic levelized cost of electricity. *Renew. Sustain. Energy Rev.* **2011**, *15*, 4470–4482. [CrossRef]
37. Gassar, A.A.A.; Cha, S.H. Review of geographic information systems-based rooftop solar photovoltaic potential estimation approaches at urban scales. *Appl. Energy* **2021**, *291*, 116817. [CrossRef]
38. Regione Autonoma della Sardegna. Database Geotopografico della Sardegna. Available online: https://www.sardegnageoportale.it/areetematiche/databasegeotopografico/ (accessed on 13 November 2022).
39. Regione Autonoma della Sardegna. Modelli Digitali del Terreno e delle Superfici della Sardegna. Available online: https://www.sardegnageoportale.it/areetematiche/modellidigitalidielevazione/ (accessed on 13 November 2022).
40. Pfeifroth, U.; Trentmann, J.; Kothe, S.; Hollmann, R.; Werscheck, M. *EUMETSAT Satellite Application Facility on Climate Monitoring; Validation Report: Meteosat Solar Surface Radiation and Effective Cloud Albedo Climate Data Record SARAH-2. 1 Climate Data Records*; EUMETSAT: Darmstadt, Germany, 2019.
41. Amillo, A.M.G.; Taylor, N.; Fernandez, A.M.M.; Dunlop, E.D.; Mavrogiorgios, P.; Fahl, F.; Arcaro, G.; Pinedo, I. Adapting PVGIS to trends in climate, technology and user needs. In Proceedings of the 38th European Photovoltaic Solar Energy Conference and Exhibition, Online, 6–10 September 2021.
42. Horn, B.K. Hill shading and the reflectance map. *Proc. IEEE* **1981**, *69*, 14–47. [CrossRef]
43. GDAL/OGR Contributors. *GDAL/OGR Geospatial Data Abstraction Software Library*; Open Source Geospatial Foundation: Chicago, IL, USA, 2022. [CrossRef]
44. Bresenham, J.E. Algorithm for computer control of a digital plotter. *IBM Syst. J.* **1965**, *4*, 25–30. [CrossRef]
45. Loutzenhiser, P.; Manz, H.; Felsmann, C.; Strachan, P.; Frank, T.; Maxwell, G. Empirical validation of models to compute solar irradiance on inclined surfaces for building energy simulation. *Sol. Energy* **2007**, *81*, 254–267. [CrossRef]
46. Faiman, D. Assessing the outdoor operating temperature of photovoltaic modules. *Prog. Photovoltaics Res. Appl.* **2008**, *16*, 307–315. [CrossRef]
47. Koehl, M.; Heck, M.; Wiesmeier, S.; Wirth, J. Modeling of the nominal operating cell temperature based on outdoor weathering. *Sol. Energy Mater. Sol. Cells* **2011**, *95*, 1638–1646. [CrossRef]
48. Huld, T.; Jäger Waldau, A.; Ossenbrink, H.; Szabo, S.; Dunlop, E.; Taylor, N. *Cost Maps for Unsubsidised Photovoltaic Electricity*; European Commission: Brussels, Belgium, 2014.
49. Lazzeroni, P.; Moretti, F.; Stirano, F. Economic potential of PV for Italian residential end-users. *Energy* **2020**, *200*, 117508. [CrossRef]
50. Terna. Lo Storico dei Dati Statistici Sull'energia Elettrica e l'Ultimo Bilancio Elettrico. Available online: https://www.terna.it/it/sistema-elettrico/statistiche/pubblicazioni-statistiche (accessed on 4 November 2022).
51. Italian Regulatory Authority for Energy, Networks and Environment (ARERA). Andamento del Prezzo dell'Energia Elettrica per il Consumatore Domestico Tipo in Maggior Tutela. Available online: https://www.arera.it/it/dati/eep35.htm (accessed on 13 November 2022).

applied sciences

MDPI

Article

Fire Risk Probability Mapping Using Machine Learning Tools and Multi-Criteria Decision Analysis in the GIS Environment: A Case Study in the National Park Forest Dadia-Lefkimi-Soufli, Greece

Yannis Maniatis [1,*], Athanasios Doganis [2] and Minas Chatzigeorgiadis [3]

[1] Department of Digital Systems, University of Piraeus, 18534 Piraeus, Greece
[2] Terra Mapping the Globe S.A., 15561 Xolargos, Greece; thanos@terra.gr
[3] Department of Digital Systems, Graduate Research Assistant at University of Piraeus, 18534 Piraeus, Greece; minas.ch32@gmail.com
* Correspondence: maniatis@unipi.gr

Abstract: Fire risk will increase in the upcoming years due to climate change. In this context, GIS analysis for fire risk mapping is an important tool to identify high risk areas and allocate resources. In the present study, we aimed to create a fire risk estimation model that incorporates recent land cover changes, along with other important risk factors. As a study area, we selected Dadia-Lefkimi-Soufli National Forest Park and the surrounding area since it is one of the most important protected areas in Greece. The area selected for the case study is a typical Mediterranean landscape. As a result, the outcome model is generic and can be applied to other areas. In order to incorporate land cover changes in our model, we used a support vector machine (SVM) algorithm to classify a satellite image captured in September 2021 and an image of the same period two years ago to obtain comparable results. Next, two fire risk maps were created with a combination of land cover and six other factors, using the analytic hierarchy process (AHP) on a GIS platform. The results of our model revealed noticeable clusters of extreme high risk areas, while the overall fire risk in the National Park Forest of Dadia-Lefkimi-Soufli was classified as high. The wildfires of 1st October 2020 and 9th July 2021 confirmed our model and contributed to quantification of their impact on fire risk due to land cover change.

Keywords: wildfire; fire risk; model; MCDA; AHP; Natura; protected zones; GIS; SVM; land cover change

Citation: Maniatis, Y.; Doganis, A.; Chatzigeorgiadis, M. Fire Risk Probability Mapping Using Machine Learning Tools and Multi-Criteria Decision Analysis in the GIS Environment: A Case Study in the National Park Forest Dadia-Lefkimi-Soufli, Greece. *Appl. Sci.* **2022**, *12*, 2938. https://doi.org/10.3390/app12062938

Academic Editor: Guijun Bi

Received: 15 February 2022
Accepted: 11 March 2022
Published: 13 March 2022

1. Introduction

The frequency of forest fires is rapidly increasing in southern Europe, posing major challenges for Greece, Italy, Portugal, Spain, and France [1]. Wildfires can represent a serious threat to human health and infrastructure, as well as ecosystems and biodiversity [2]. More specifically, the impact of wildfires on human health can be either direct, causing severe physical damage due to burns, or indirect, since the exposure to pollutants such as ozone and PM [3] can lead to serious disorders. In addition, large wildfires can damage properties or critical infrastructure, such as electricity grids and houses, resulting in major economic losses [4]. Finally, wildfires play an important role in ecological balance, in which humans are a part of. The increase in fire frequency in the past few years enhances forest degradation and biodiversity loss [5].

In terms of biodiversity, Greece is one of the richest countries in Europe, having the highest number of flora species among the Balkan countries. In fact, it contains over 5700 different species of flora, 20% of which are endemic to the country. Most of these species are located in the northern regions of the country, thanks to the ideal geographic and climate conditions [6,7]. It seems that species richness of Greece combined with the high risk of fire in the Mediterranean region—because of its hot and dry summers—makes Greece extremely susceptible to wildfires. Consequently, the average annual burned area caused by wildfires has shown an increasing trend in the past decades. During 2001–2017, wildfires burned an average of 55,000 ha per year, most of which was covered by forested areas [8]. Recently, one of the most disastrous wildfires took place in August 2021 in Evia island, where 34,893.5 ha of forest and 1111.6 ha containing houses and infrastructure were destroyed or seriously affected by fire [9]. As a result, it is critical to study how various factors influence the probability of fire occurrence, in order to create a fire risk layer for the Fire Management Geographic Information

Fire risk expresses the likelihood of a fire occurring during a specific time period and place. The risk is the result of the different hazardous parameters interacting with the conditions of vulnerability, which are present in the region [10]. On the one hand, hazardous parameters describe the danger of fire occurrence and on the other hand, vulnerability expresses the predilection of an area to be negatively affected by wildfire [11]. It is very common for the terms 'fire risk' and 'fire danger' to have interchangeable meanings. The factors that influence the ignition and development of fires constitute the fire danger. Fire ignition can derive from natural causes (mostly thunder), or it can be a result of human activity [12]. According to a study, approximately 93% of fires in Northern Europe are caused by humans, either intentionally or unintentionally [13], and thus the location of populated places and roadways is critical in identifying areas at high risk of fire. The development of fires is influenced by topography, meteorological conditions, fuel condition, and fuel availability [14]. Many studies have shown that vegetation and topography are the key elements responsible for fire severity in many types of forests. [15–18]. Since topographic features influence the distribution of local climate, topography is an important factor in fire propagation. Fires spread quickly across steep and upward slopes, but slowly in places with a downhill slope [19]. Moreover, the probability of fire occurrence may vary in different elevations on the basis of factors such as temperature and vegetation [20]. The topographic wetness index (TWI) [21] is another parameter that contributes to fire spread and ignition. To a certain extent, TWI simulates the impact of topography to soil and fuel moisture [22,23].

Geographic information systems (GIS) is mature technology and effective platform to analyze, visualize, and disseminate spatial and temporal data and information. GIS is a multidisciplinary approach that can combine methods from science, engineering, and the economy with the experience of field officers to produce robust knowledge in firefighting. GIS, besides its analytical capabilities, is the ideal platform for coordination, information exchange, and awareness provision for all involved stakeholders (all levels of authorities, fire department, police, forest services, agricultural coops, citizens, etc.).

Multiple levels of spatial and nonspatial data and information related to fire risk, such as meteorological data, land cover, vegetation features, and topography, in the form of historical information is combined and evaluated to create detailed fire risk maps [24–26]. The information to be used by the fire risk model has to be reliable, the most recent, easy to obtain, and processable with reasonable H/W and S/W resources in order to produce an update fire risk map. The key for the analysis is the determination and assignment of the proper weights between all these pieces of information. Many studies, in particular, have used multi-criteria decision analysis (MCDA) in conjunction with the analytic hierarchy process (AHP), which assigns weights to the influencing parameters, so as to successfully develop fire risk maps [14,24–28]. In the AHP framework, a decision is broken down into a hierarchy of criteria or alternatives, and subsequently one can evaluate the significance of each criterion to the final decision, given the relevant weights between each pair of criteria [29]. After an exhaustive review of the bibliography, we proposed seven factors to be used as criteria in AHP to estimate fire risk: the land cover (LC), the elevation, the aspect, the slope, the TWI, and the distance from roads (DfR) and settlements (DfS). Out of those seven factors, studies have shown that land cover is the most important for estimating fire risk [24,27], especially in cases where the land cover indicates the type of vegetation that covers the area [28,30]. It is evident that having a detailed and updated depiction of land cover is critical for estimating fire risk. In view of this fact, GIS can be used to classify land cover and vegetation from satellite imagery with the implementation of machine learning algorithms [31,32]. Various algorithms have been used by different studies for land classification, such as k-means clustering [33], maximum likelihood classification [34], and support-vector machines (SVM) [35]. In fact, support-vector machine models have been used for land classification with promising results [31,35].

In this paper, the SVM algorithm was applied to satellite images obtained in 2019 and 2021 in order to create detailed land cover maps for the Natura 2000 (GR1110005) zone, which includes the Dadia-Lefkimi-Soufli National Park Forest in the county of Evros. Subsequently, the use of these land cover maps in combination with six other important fire risk factors can determine the fire risk of the National Park for September 2019 and September 2021. On the basis of the fire risk maps of 2019 and 2021 in conjunction with the burned areas from the past fires of 1st October 2020 and 9th July 2021, we evaluated how substantial land cover changes can affect fire risk mapping. In order to make a valid comparison, we chose to include in the fire risk model factors that remain relatively constant for long periods of time. This creates a baseline fire risk map of our study area.

2. Study Area

Our study area is the Natura 2000 zone with codename GR1110005, which coincides with the National Park Forest of Dadia-Lefkimi-Soufli. The study area is located in Evros county, as shown in Figure 1, and it extends from 26.03° E to 26.32° E and from 40.98° N to 41.26° N, covering a total area of 42,481 ha. The climate in the area is Mediterranean, with daytime maximum average temperatures of 32 °C in August and lowest average temperature of 8 °C in January [36]. The average number of rainy days per year is 13.3 and the average yearly rainfall is 732 mm [37]. The lowest point of the study area has a height of 10 m, and the highest point is located in Kapsalo at 620 m [37,38].

The National Park Forest of Dadia-Lefkimi-Soufli contains two protected zones, A1 and A2, which cover an area of 7350 ha. Oak and pine trees make up the majority of the forested areas. The spatial distribution of the different types of trees in the National Park can be divided into two areas. The center is covered with pine trees, whereas the north and southwest are covered mostly with oak trees [39]. During 2019 and 2021, two major fires occurred within the National Park, both of which took place in the southern region, on the northern part of the village Lefkimi. The first burstfire occurred on 1st October 2020, burning approximately 694 ha, and the second one took place on 9th July 2021, burning approximately 242 ha.

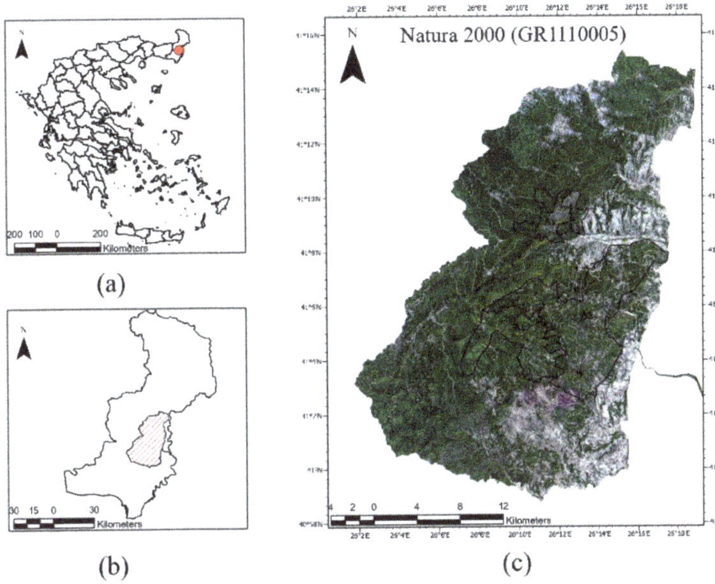

Figure 1. (**a**) Location of the study area in Greece (**b**) and in Evros county. (**c**) Longitude and latitude of Natura 2000 area (GR1110005) that coincides with the National Park Forest of Dadia-Lefkimi-Soufli, including the protected zones A1 and A2.

3. Data and Methods

To construct the land cover maps, Sentinel-2 images were used, which were captured on 18th September 2019 at 09:06 a.m. and on 27th September 2021 at 09:06 a.m., with minimal cloud coverage (<0.1%) [40]. Both images were level 2 Sentinel-2 products, and thus they had already received atmospheric correction. Next, the spectral bands B03, B04, and B08 were extracted from the original images, with spatial resolution of 10 m, in order to produce the color infrared images needed for the land classification. Finally, to classify the color infrared image, we used a supervised machine learning (SML) model, using the application of the support vector machine (SVM) algorithm. SVM algorithms have been proven to be a reliable method of creating land cover maps from Sentinel-2 images [31,32,35].

For the calculation of the topographic factors, the freely available digital elevation model (DEM) of the Copernicus Land Monitoring Service was used. The Copernicus DEM offers spatial resolution of 25 $\times$ 25 m, with vertical accuracy of $\pm$7 m (RMSE). For the purpose of our study, the 1000 $\times$ 1000 km tile with codename E50N20 was used, and the elevation of our study area was isolated from it [38]. The remaining topographic factors (slope, aspect, and TWI) were derived by analyzing the DEM. The roads and settlement locations were downloaded from open data sources [41,42]. For the raster analysis and calculations mentioned above, the GIS software ArcGIS Pro 2.9.1. was used.

In order to validate our results and examine the impact of the fire on land cover and subsequently the fire risk, detailed burn scar maps from the National Management Body of Dadia-Lefkimi-Soufli Forest National Park were used [37], which depict the extent of fires that occurred during October 2020 [43] and July 2021 [44]. VIIRS hotspot locations were also used to validate the spatial extend of the burned areas [45]. The workflow of our method is represented in Figure 2.

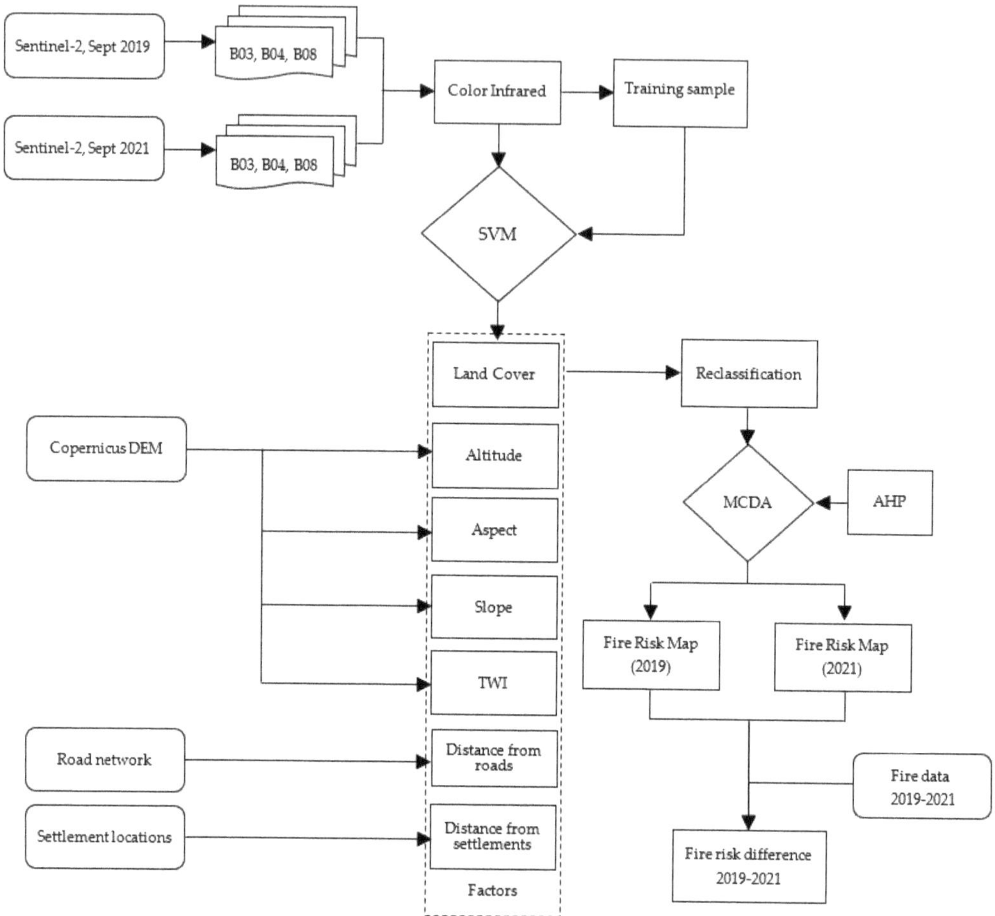

Figure 2. Flowchart of the applied methodology.

3.1. Factors

3.1.1. Land Cover (LC)

To create the land cover maps, we used an SVM algorithm to process color infrared images from September 2021 and September 2019, which were derived from the combination of Sentinel-2 B03, B04, and B08 bands. The resulting color infrared images are presented in Figure 3, in which vegetation appears in shades of red, bare land in cyan or white, and water in black.

Color infrared images can help distinguish among different plant types, depending on their leaf characteristics [46]. Inside the National Park of Dadia-Lefkimi-Soufli, oak and pine trees account for more than the 70% of vegetation [39]. Oak trees belong to the broad-leaved tree family, and therefore they appear in brighter red in the color infrared image. In Figure 3, oak trees can be seen as clusters of bright red in the northern and southwestern parts of the national park. Pine trees have thinner leaves, and thus they appear to have a darker red color. Finally, shrubs and low grass appear in faint red.

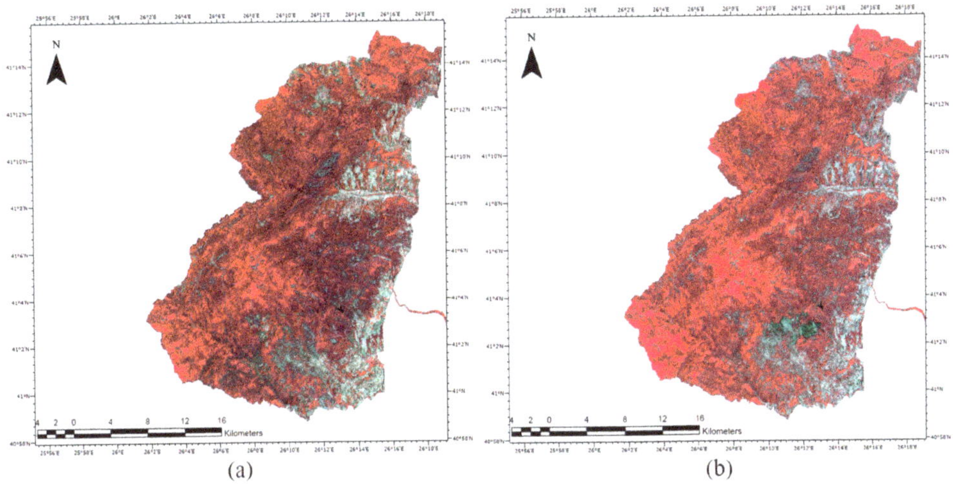

Figure 3. Color infrared image of the study area produced by the combination of spectral bands B03, B04, and B08 (**a**) from September 2019 and (**b**) from September 2021.

To classify the images, we considered six classes: pine forest, oak forest, shrubs and low grass, bare land, water bodies, and built-up areas. To train the SVM algorithm, we carefully gathered multiple samples of homogeneous parts from each image, representing one of the six classes. After the classification of the image, we made some adjustments to the product image, mainly to distinguish some parts of bare land from the buildup areas. The accuracy of the land cover classification was estimated using the Kappa coefficient, which was found to be 0.87. The final land cover maps from September 2019 and September 2021 of the National Park of Dadia-Lefkimi-Soufli are presented in Figure 4a,b, respectively.

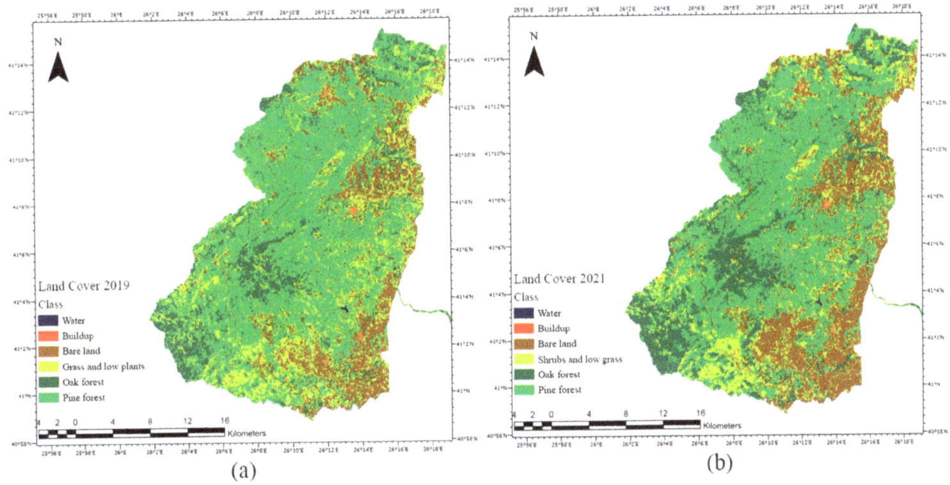

Figure 4. Final land cover map produced by the SMV algorithm (**a**) from September 2019 and (**b**) from September 2021.

Each type of tree has different flammability properties. Considering that pine trees are more flammable than oak trees, the forest areas were classified accordingly [47]. Finally, since water bodies cannot ignite, they were classified with the fire risk class 'no risk'.

The classification of land cover, based on the fire risk, is shown in Table 1, and the final reclassified risk map of the land cover maps is presented in Figure 5.

Table 1. Fire risk classification of land cover.

Land Cover Class	Risk Class	Risk Description
Pine forest	5	Extremely high
Oak forest	4	High
Shrubs and low grass	3	Medium
Bare land	2	Low
Buildup	1	Extremely low
Water body	0	No risk

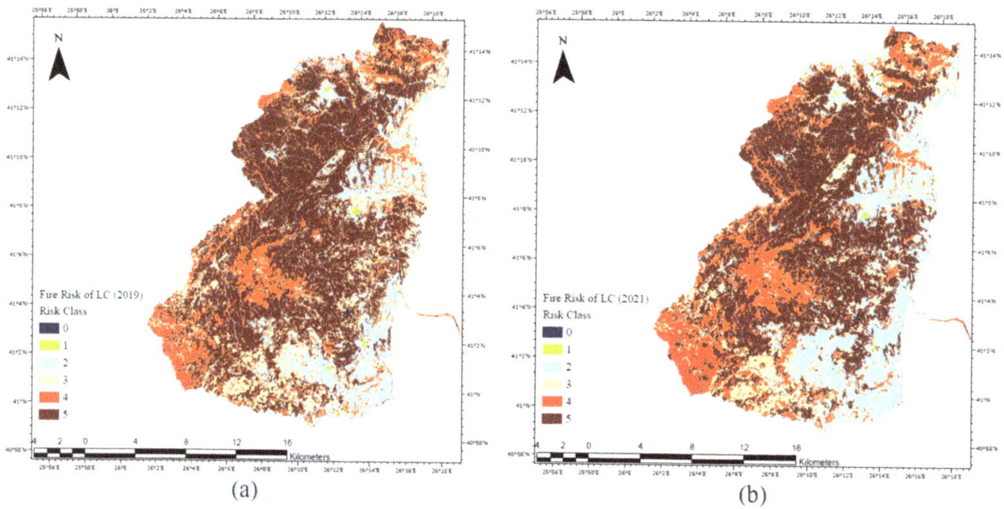

Figure 5. The fire risk classification of land cover (**a**) for September 2019 and (**b**) for September 2021.

3.1.2. Altitude

Altitude influences the humidity of vegetation and temperature. Vegetation in high altitudes has higher rates of humidity and lower temperature [31]. Moreover, high altitudes usually have lower vegetation density. Considering the topographic characteristics of the area, we distributed the fire risk into five classes, as shown in Table 2. Elevation in our study area, according to DEM [38], ranged from 10 m meters to 645 m meters. The altitude raster and the final reclassification of fire risk appear in Figure 6.

Table 2. Fire risk classification of altitude.

Altitude (m)	Risk Class	Risk Description
10–100	5	Extremely high
100–200	4	High
200–300	3	Medium
300–400	2	Low
>400	1	Extremely low

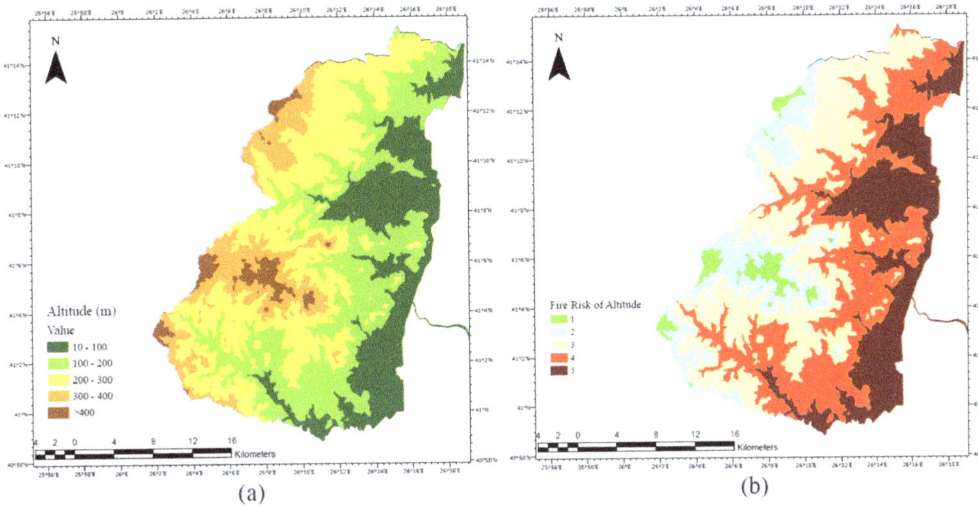

Figure 6. (**a**) Altitude. (**b**) Fire risk classification of altitude.

3.1.3. Aspect

In the northern hemisphere, south-oriented slopes receive more sunlight, and thus the vegetation loses humidity faster and becomes more flammable [28]. Moreover, because of the difference in sunlight distribution among the different orientations of slope, the southern aspects usually have more dense vegetation. Less humidity and dense vegetation results in higher fire risk, and thus vegetation facing south is more flammable. The fire risk classification of the aspect appears in Table 3 [25,28].

Table 3. Fire risk classification of aspect.

Aspect	Risk Class	Risk Description
South	5	Extremely high
Southeast–East	4	High
Northeast	3	Medium
North	2	Low
Flat–Southwest–West–Northwest	1	Extremely low

The aspect derives from the DEM and the is presented with different colors depending on the orientations. The final aspect raster along with reclassified fire risk map can be seen in Figure 7.

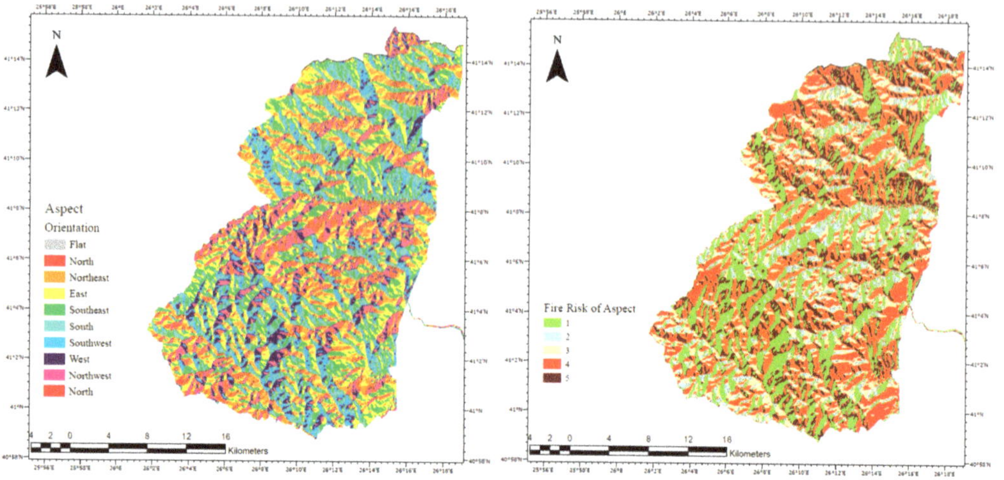

(a) (b)

Figure 7. (**a**) Aspect. (**b**) Fire risk classification of aspect.

3.1.4. Slope

Fire propagates faster on steeper slopes because the flames can reach higher vegetation more easily at great surface angles [30]. Moreover, on steep slopes, water runoff increases, resulting in less soil moisture [24]. Both of these factors make areas with steeper slopes have a higher risk of fire. We derived the fire risk classification of slope as shown in Table 4 [25]. In order to calculate the slope raster, we used the DEM and chose to present the results in percentage. The slope raster of the National Park Forest of Dadia-Lefkimi-Soufli along with the fire risk map of the slope is presented in Figure 8.

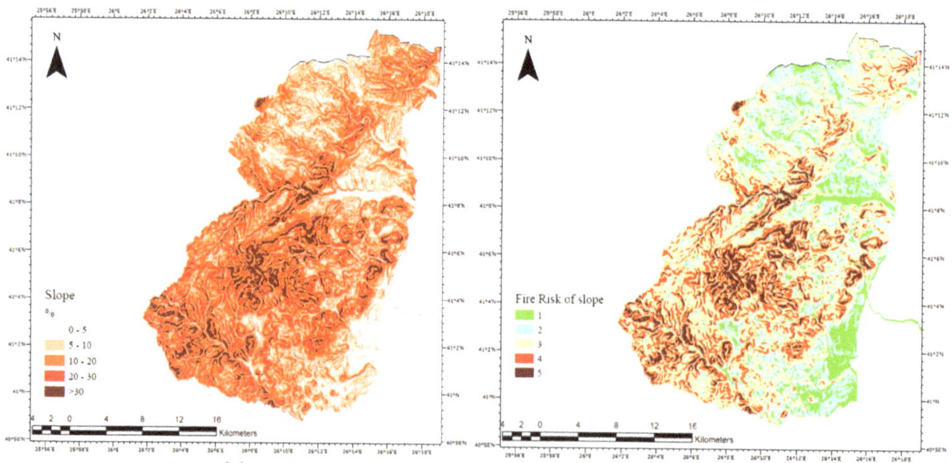

(a) (b)

Figure 8. (**a**) Slope. (**b**) Fire risk classification of slope.

Table 4. Fire risk classification of slope.

Slope (%)	Risk Class	Risk Description
>30	5	Extremely high
20–30	4	High
10–20	3	Medium
5–10	2	Low
0–5	1	Extremely low

3.1.5. Topographic Wetness Index (TWI)

The TWI can simulate water concentration can be derived from topography. The presence of water affects soil moisture and makes the surrounding vegetation harder to ignite [22]. We calculated the TWI of the study area from the total catchment area, the flow width, and slope from the DEM [21]. The risk classification of the TWI is presented in Table 5. The TWI raster of the study area along with the fire risk map of TWI is shown in Figure 9.

Table 5. Fire risk classification of TWI.

TWI	Risk Class	Risk Description
4–6	5	Extremely high
6–7	4	High
7–8	3	Medium
8–9	2	Low
>9	1	Extremely low

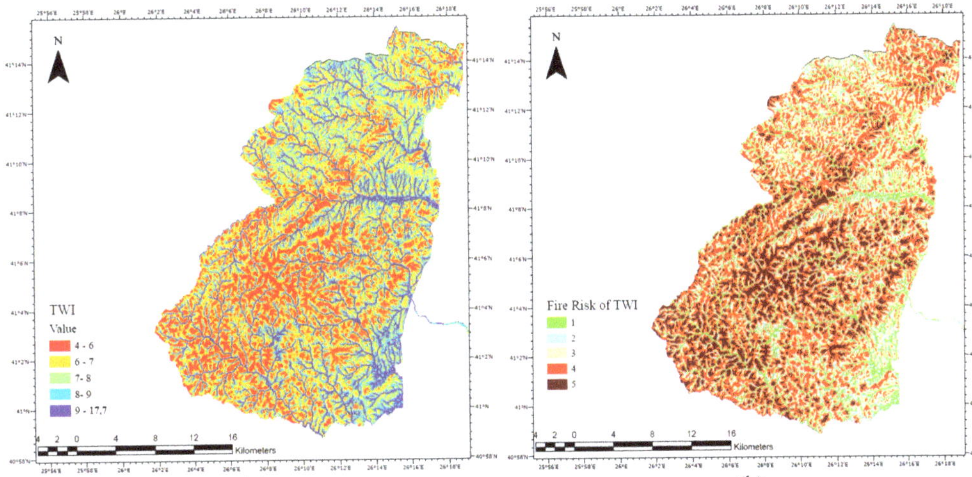

(a) (b)

Figure 9. (**a**) TWI. (**b**) Fire risk classification of TWI.

3.1.6. Distance from Roads

Human activities near roads can be the cause of fire ignition, and therefore the areas surrounding the road network are at a higher fire risk [48]. To attribute fire risk to those areas, we took into consideration previous studies along with the structure of the road network [24,28]. The first 200 m near the road network was determined to be at high risk of fire, and afterwards the risk decreased by one class at 200 m intervals. The classification of fire risk is presented in Table 6. Multi-buffer rings of 200 m each were calculated around each road segment and transformed to rasters in order to be incorporated into the model (Figure 10).

Table 6. Fire risk classification of the area around the road network.

DfR (m)	Risk Class	Risk Description
0–200	5	Extremely high
200–400	4	High
400–600	3	Medium
600–800	2	Low
>800	1	Extremely low

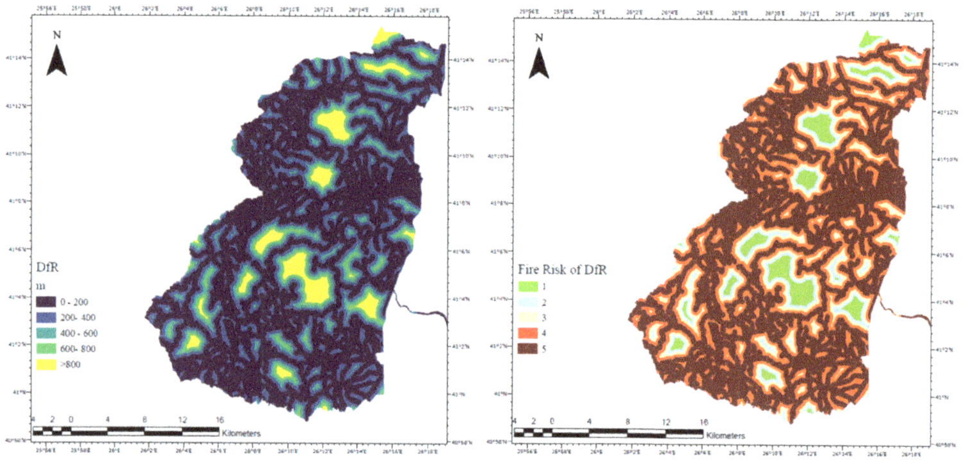

(a) (b)

Figure 10. (**a**) Rasterized buffer zones every 200 m from the road network. (**b**) Fire risk classification based on the road network.

3.1.7. Distance from Settlements

The distance around settlement locations affects the risk of fire similarly to that of the road network. The areas closer to settlements are in higher risk than those farther away [13]. To distribute the fire risk, we took the spatial extent of the settlements into consideration. Since most of the settlements inside and near our study are small, we estimated that their average extent is 500 m. Taking this into account, we assigned the area inside a radius of 900 m around the settlements to be in extreme risk of fire, and afterwards the risk decreased by one class at 400 m intervals, as it is depicted in Table 7. Buffer zones using the aforementioned distances were applied around each settlement. The results and the assigned fire risk is shown in Figure 11.

Table 7. Fire risk classification of the area around settlement locations.

DfS (m)	Risk Class	Risk Description
0–900	5	Extremely high
900–1300	4	High
1300–1700	3	Medium
1700–2100	2	Low
>2100	1	Extremely low

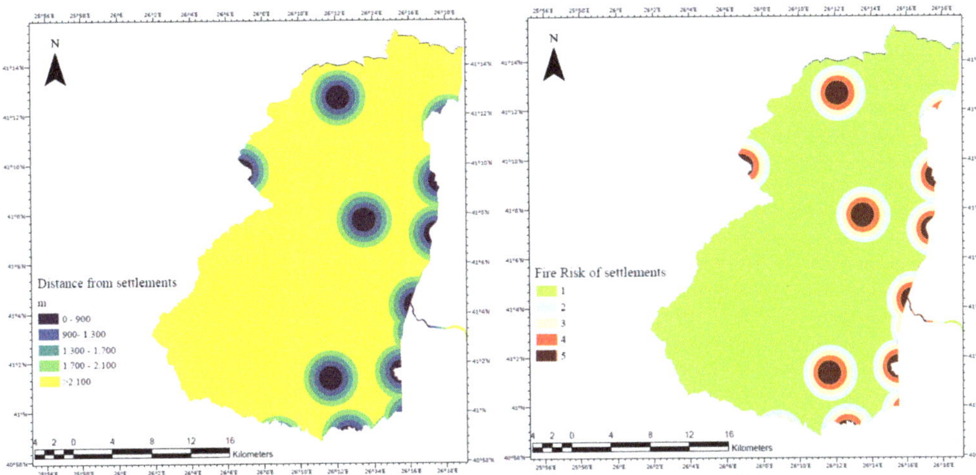

(a) (b)

Figure 11. (**a**) Rasterized buffer zones of the distance from settlements. (**b**) Fire risk classification of the area around settlement locations.

3.2. Attribution of Weight to the Factors

To calculate the weight of each factor, we used the AHP as a multi-criteria method. The AHP can estimate the significance of each factor given the pairwise comparisons among each one of the seven factors [29]. In our model, each factor belongs to one major category that influences fire risk. Particularly, slope, DEM, aspect, and TWI reflect the impact of topography and to some extent fuel condition on fire risk. The distance from roads and settlements captures the impact of human activity on fire ignition. Finally, land cover depicts the alternations of fire risk, due to fuel availability and fuel type. In order to estimate the pairwise comparisons, we consulted past studies and expert opinions. We considered the land cover to be the most important factor for this estimation [24,25,27]. Moreover, human activity plays a key role in fire risk identification, since in most cases, humans are the main cause of fires [13]. Therefore, the distance from roads and settlements has a serious impact on fire risk. The most important topographic factor is TWI, since it has a direct correlation with soil humidity, whereas the rest topographic factors contribute less to the overall fire risk. The final distribution of pairwise comparisons is presented in Table 8. Afterwards, the weights were calculated using the mathematical procedure established by Thomas L. Saaty [49]. The final weights of each factor are also shown in Table 8.

Table 8. Pairwise comparisons of fire risk factors along with the assigned weight.

	Land Cover	Altitude	Aspect	Slope	TWI	DfR	DfS	Weight
Land cover	1	3	3	3	3	2	2	0.27
Altitude	0.33	1	3	2	0.5	0.33	0.33	0.09
Aspect	0.33	0.33	1	0.5	0.25	0.33	0.33	0.05
Slope	0.33	0.5	2	1	0.5	0.33	0.33	0.07
TWI	0.33	2	4	2	1	0.33	0.33	0.12
DfR	0.5	3	3	3	3	1	3	0.23
DfS	0.5	3	3	3	3	0.33	1	0.17
SUM	3.32	12.83	19	14.5	11.25	4.65	7.32	1

To verify the consistency of our comparison estimations, we calculated the consistency ratio (CR) by applying the following equations [30],

$$CI = (\lambda max - n)/(n - 1) \tag{1}$$

$$CR = CI/RI \tag{2}$$

The λmax in Equation (1) is the perturbated eigenvalue of the matrix constructed by the pairwise comparisons, as depicted in Table 8. n is the order of the matrix, n = 7. The consistency index (CI) in Equation (1) measures the difference between λmax and the exact eigenvalue, n. The CR in Equation (2) is calculated from the random consistency index (RI) [50], as shown in Table 9.

Table 9. Values of the random consistency index (RI).

n	1	2	3	4	5	6	7
RI	0	0	0.58	0.9	1.12	1.24	1.32

According to Table 9, RI = 1.32 for seven factors. Subsequently, concerning our pairwise estimations, CR = 0.07. Since CR < 0.1, the estimations of the pairwise matrix were consistent.

The fire risk maps were calculated by the weighted sum of all factors

$$\text{Fire Risk} = 0.27 * LC + 0.09 * \text{Altitude} + 0.05 * \text{Aspect} + 0.07 * \text{Slope} + 0.12 * TWI + 0.23 * DfR + 0.17 * DfS \tag{3}$$

4. Results

The fire risk maps for September 2019 and September 2021, with spatial resolution of 25 m × 25 m, were calculated using Equation (3), and they are presented in Figure 12, along with the relative fire risk scale value. The risk class for most areas was unchanged, which is expected. The parameters used in our model remained relatively invariant for long periods of time. Therefore our risk map represents the baseline fire risk in the area [24].

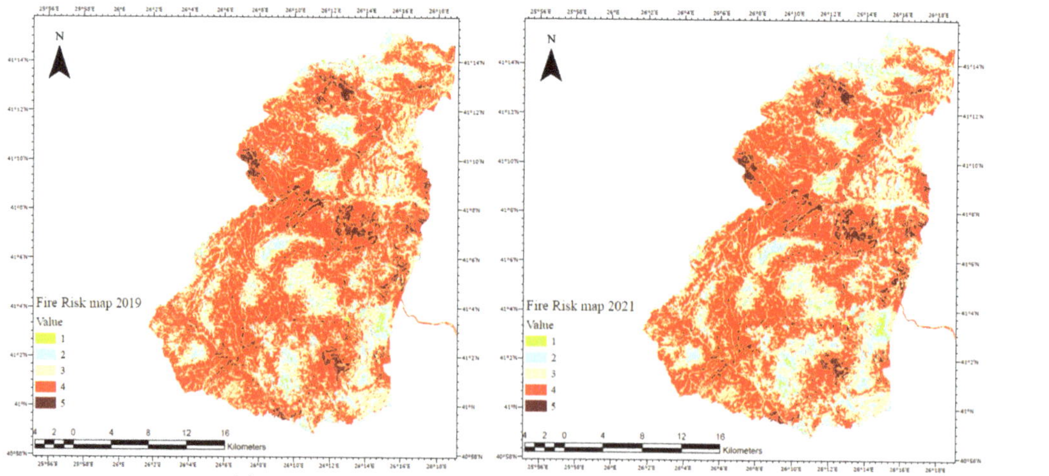

(a) (b)

Figure 12. Fire risk in the National Park Forest of Dadia-Lefkimi-Soufli (**a**) for September 2019 and (**b**) for September 2021.

The average fire risk in the National Park Forest of Dadia-Lefkimi-Soufli was high for both years. Specifically, 50% of the total 42,481 ha of our study area for 2019 was considered to be at high fire risk. Similarly, 48% of the total area was considered to be at high fire risk for 2021. Moreover, for both years, 5% of the National Park was at extremely high risk of fire.

The extreme fire risk areas form clusters. The most noticeable ones are located near the center of the National Park, spreading along the line that connects the points with coordinates 41°10′ N, 26°7′ E and 41°6′ N, 26°16′ E. Two additional extremely high risk areas were detected. The first one is located in the northwest of the National Park, near the settlement of Giannouli, and the second one is located in the southeast of the park, near the settlement of Lefkimi. The overall distribution of fire risk for September 2019 and September 2021 in the National Park is presented in Table 10.

Table 10. The distribution of fire risk in the National Park of Dadia-Lefkimi-Soufli for September 2019 and September 2021.

Risk Class	Risk Description	Fire Risk Areas (Sept 2019)	Fire Risk Areas (Sept 2021)
5	Extremely high	5%	5%
4	High	50%	48%
3	Medium	33%	34%
2	Low	11%	12%
1	Extremely low	1%	1%

Within the National Park Forest of Dadia-Lefkimi-Soufli, two major fire incidents took place between September 2019 and September 2021. The first one was recorded in October 2020, and the second one in July 2021 [46]. In order to validate our results and measure the impact of change in the land cover on the fire risk, due to the fires, each fire incident was examined separately as follows.

4.1. Impact of Fire in October 2020

The fire of October 2020, as stated by the fire department, started in the north of the village of Lefkimi, near the southwest extreme high fire risk area, and burned approximately 694 ha. According to VIIRS hotspot measurements, the brightness temperature during the fire ranged from 24 °C to 81 °C [46]. Moreover, the area affected by the fire before the fire occurrence was considered at high risk. In fact, according to our fire risk map of 2019, 41% of the area was classified as high risk and 36% as medium risk. The fire extent and the fire risk map of 2019, along with the distribution of the fire risk inside the affected area, are presented in Figure 13.

The fire had a significant impact on the fire risk in the area. The change in land cover we identified with the SVM algorithm passed on the fire risk and was captured by the difference among the fire risk maps inside the extent of the affected area before and after the fire incident. The fire risk of the area dropped from high to medium–low risk. The high-risk areas dropped from 41% before the fire to 10% after the fire, while the low-risk areas increased by 17%. These changes are attributed to the loss of vegetation from the fire on 5 October 2020. The fire extent related to the fire risk map of 2021, along with the updated distribution of the fire risk inside the affected area, is presented in Figure 14.

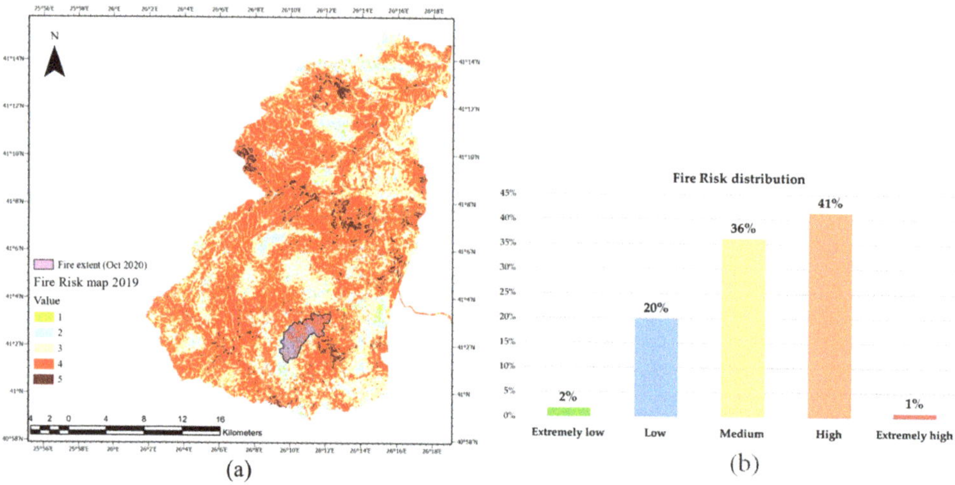

Figure 13. (**a**) The extent of the fire on 5 October 2020 relative to the fire risk map of 2019. (**b**) The fire risk distribution inside the affected area before the fire on 5 October 2020.

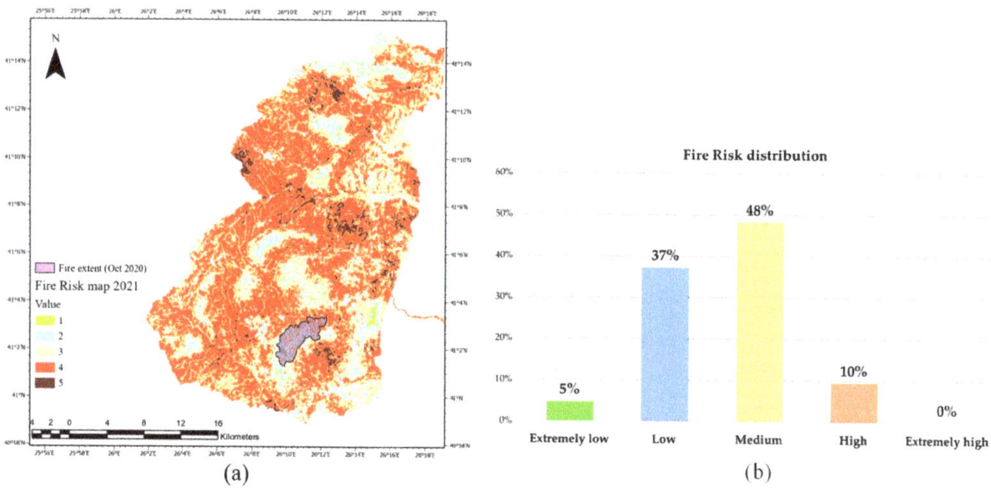

Figure 14. (**a**) The extent of the fire on 5 October 2020 relative to the fire risk map of 2021. (**b**) The updated fire risk distribution inside the affected area, after the fire on 5 October 2020.

4.2. Impact of Fire in July 2021

The second serious fire incident in the National Park Forest of Dadia-Lefkimi-Soufli was recorded on 9th July 2021, in the north of the settlement of Lefkimi. The fire burned approximately 242 ha of forested areas, and it was close to the limits of the fire that occurred in October 2020, as is shown in Figures 14 and 15. According to VIIRS hotspot measurements, the brightness temperature during the fire ranged from 59 °C to 32 °C [46]. According to our model, the affected area before the fire occurrence was considered to be at high risk of fire. In particular, 59% of the overall area was classified as having a high fire risk, and 10% as having an extremely high fire risk. On the contrary, only 4% of the area was classified as low and 0% as extremely low fire risk. The detailed distribution of the fire risk inside the affected area (according to the risk map of 2019) is presented in Figure 15.

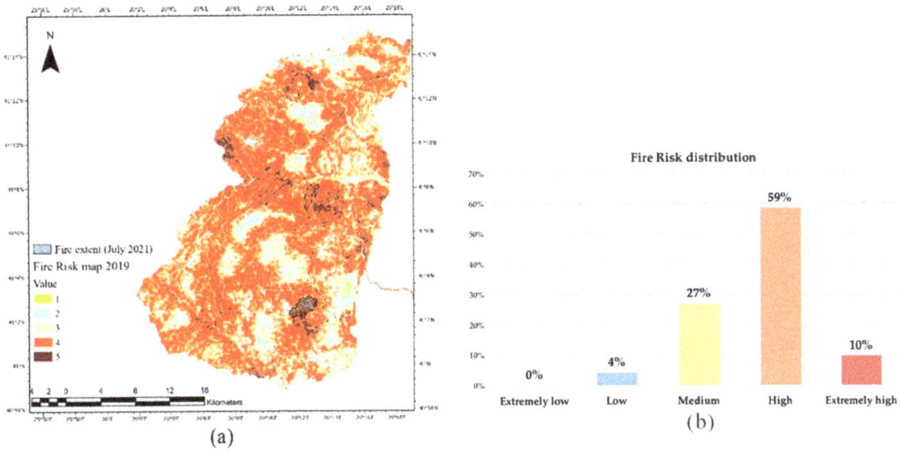

Figure 15. (**a**) The extent of the fire on 9 July 2021 relative to the fire risk map of September 2019. (**b**) The fire risk distribution inside the affected area, before the fire on 9 July 2021.

The fire had a significant impact on the overall fire risk classification of the area. The fire risk of the affected area after the fire was classified as medium with 49%. After the fire incident, the high-risk areas dropped from 59% of the whole area to 26%, while the extremely high risk areas dropped from 10% to 3%. On the other hand, areas categorized as low risk increased by 18%. The fire extent related to the fire risk map of September 2021, along with the updated distribution of the fire risk inside the affected area, is presented in Figure 16.

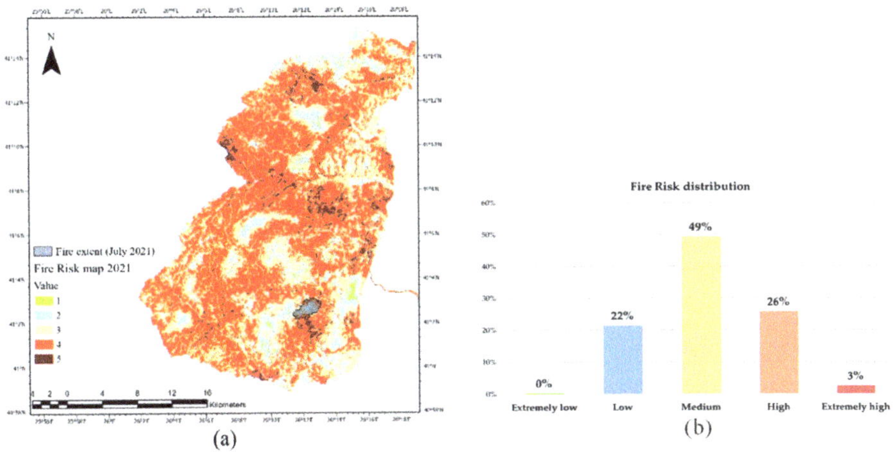

Figure 16. (**a**) The extent of the fire on 9 July 2021 relative to the fire risk map of 2021. (**b**) The updated fire risk distribution inside the affected area, after the fire on 9 July 2021.

5. Discussion

In this study, we combined elevation, slope, aspect, TWI, land cover, settlement location, and road networks to create a fire risk model. Subsequently, we applied the model to the National Park Forest of Dadia-Lefkimi-Soufli. Topography, vegetation, and human activity are the major categories represented in our model that affect wildfire generation and spread. Similar factors have been used by other studies, in combination with AHP to assign weights, in order to calculate fire risk in various areas [14,24–28,30]. In our analysis, land cover was proved to be the factor with the highest weight. This does not mean that topography and human activities are of less importance. In fact, human activity is the main cause of wildfire ignition. Approximately 93% of fires in Northern Europe are caused by humans, either intentionally or unintentionally [13]. Land cover, besides having the highest weight in the resulting model, also changes more frequently compared to human activities (road network, settlements) and topography. For this reason, we derived land cover classification from Sentinel 2 imagery since they are georeferenced, frequently updated, freely available, and have suitable spatial resolution. Classification was carried out with the use of SVM algorithm, since it has already been used in similar applications with promising results [31,35].

With the combination of AHP (to determine the fire risk) and the SVM algorithm (to classify the land cover), we managed to identify the baseline fire risk of the National Park Forest of Dadia-Lefkimi-Soufli for September 2019 and September 2021. According to those fire risk maps, most of the areas in the National Park Forest of Dadia-Lefkimi-Soufli are classified as high risk. More specifically, the map of 2021 reveals that 5% out of the total area of the National Park was classified as extremely high risk, and 48% was high risk. Consequently, fire risk distribution in the National Park suggests that local authorities should be at high alert, especially during heatwaves and near the areas classified as extreme high risk.

We also examined in detail the impact of the land cover change on fire risk in the areas affected by the two major fire incidents (October 2020 and July 2021). It is concluded that the average risk of those areas dropped significantly, while the rest of the fire risk map remained relatively unchanged between September 2019 and September 2021. It is evident that land cover changes caused by past fires have a significant drop on the fire risk of the affected areas. The main cause is the loss of the highly flammable pine tree forest near the settlement Lefkimi. Considering the effect of land cover changes on fire risk mapping, it is important that fire risk management plans incorporate those changes and reallocate resources accordingly on a local scale. In this way the SVM algorithm, along with other classification algorithms [33–35,51], can offer a powerful tool for updating fire risk maps year by year. Finally, it is notable to point out that both October 2020 and July 2021 fires started near areas classified as 'extreme high' fire risk. This is yet another validity indicator of the proposed fire risk model. Additionally, the above-mentioned fire incidents spread mostly at areas classified as 'high risk' (Figures 13 and 15).

We acknowledge that daily risk maps at the national [52] and European levels [53], which are mostly derived from weather data, are freely available and easily accessible to all stakeholders, including the citizens. Prediction models based on weather information alert civil protection and fire services on areas of increased readiness. Fire risk models and maps (such as the one proposed in this study) are supplementary to weather data, indicating high risk areas usually at higher spatial resolution and where proactive measures can, or should, be taken. Those measures can include the optimal allocation of observatories and/or fire service areas by using GIS tools such as visibility, network, or other suitable analysis.

We utilized GIS technology for fire risk model development, not only because of GIS's analytical and presentation capabilities, but also due to information dissemination and its integration ability along with other organizational workflows. Developing the proposed model, a key issue addressed is data availability. As the reader can see all data utilized are from reliable [38,40] and freely referenced or downloaded sources [41,42]. As a result, local authorities, which are responsible in specifying precautional policies and measures (especially in high fire risk periods), such as increased supervision, temporal road closures to traffic, prohibition of certain activities, and more optimal resource allocation, have the means to identify areas of higher risk.

The spatial resolution of the presented model is limited to the resolution (25 by 25 m) of the DEM used to calculate it [38]. Higher-resolution DEMs are calculated by mapping agencies during orthophoto production workflows. They can also be acquired by drones [54]. A question of further research and investigation is if higher resolution DEMs help in a better understanding and mapping fire risk or just add higher frequency data, which sometimes are noise rather than actual information. A constrain of our model is the lack of meteorological factors. Furthermore, it is important to note that in order to determine the fire risk of an area, meteorological factors such as temperature, wind, and humidity play a very important role [26]. These factors change dynamically, and therefore it is hard to establish a baseline risk map to compare results from different years. Human activity layers of information (road network and settlements) were derived from OpenStreetMap in testing our model. More important than the accuracy of road network is completeness and the level of update. State agencies and local authorities are advised to use the most updated information they have access to. Other data sources of human activities, such as electricity grids and landfills, which are of great importance, can be incorporated in our model, creating multi-buffer rings, similar to how road network and settlement areas were treated.

6. Conclusions

Wildfires, unfortunately, are an inevitable consequence of climate crisis. Every year, we are witnessing more and more devastating wildfires in the western United States, Amazon basin, South Europe, Siberia, Australia, and elsewhere, with a priceless impact on our environment. Understanding and modelling the phenomenon can make us more effective in addressing it. Especially in firefighting, the timely response is the most crucial factor to fight it. The role of fire risk models in conjunction with other GIS analysis tools can provide us useful information for optimal arrangement of all available resources before the ignition of the phenomenon such as selecting supervision locations for areas characterized as 'high risk' and even allocate firefighting trucks for a more immediate response in case of an incident. Knowledge of 'high risk' areas can assist all levels of administration to increase citizen awareness and take targeted proactive measures.

Models and all data needed to support it should be free and easily accessible for agencies and authorities to integrate it with their systems. All necessary data to implement the model can be easily found at Copernicus services and other European or National spatial data infrastructures. The most essential data to classify land cover, Sentinel 2 images, are available freely worldwide through the same services. We will keep working and testing the model in other areas to test its portability. Findings and suggestions will help us to improve it. All future improvements will be embedded and published.

Author Contributions: Conceptualization, A.D. and Y.M.; methodology, A.D. and Y.M.; software, A.D.; validation, M.C.; formal analysis, A.D.; investigation, A.D. and M.C.; resources, M.C. and Y.M.; data curation, M.C.; writing—original draft preparation, M.C.; writing—review and editing, A.D. and Y.M.; visualization, M.C.; supervision, A.D. and Y.M.; project administration, A.D. and Y.M.; funding acquisition, Y.M. All authors have read and agreed to the published version of the manuscript.

Funding: This research received no external funding.

Institutional Review Board Statement: Not applicable.

Informed Consent Statement: Not applicable.

Data Availability Statement: The 25 m resolution DEM data can be downloaded from Copernicus Land Monitoring Service (https://land.copernicus.eu/imagery-in-situ/eu-dem/eu-dem-v1.1, accessed on 5 October 2021). The Sentilel-2 images for land cover classification are found at Copernicus Open Access Hub (https://scihub.copernicus.eu/dhus/#/home, accessed on 10 December 2021).

Acknowledgments: We would like to thank Paraschi Eirini for copy editing the final manuscript.

Conflicts of Interest: The authors declare no conflict of interest.

References

1. Tedim, F.; Xanthopoulos, G.; Leone, V. Forest Fires in Europe. In *Wildfire Hazards, Risks and Disasters*; Elsevier: Amsterdam, The Netherlands, 2015; pp. 77–99. ISBN 978-0-12-410434-1.
2. Martin, D.; Tomida, M.; Meacham, B. Environmental Impact of Fire. *Fire Sci. Rev.* **2016**, *5*, 5. [CrossRef]
3. Dittrich, R.; McCallum, S. How to Measure the Economic Health Cost of Wildfires—A Systematic Review of the Literature for Northern America. *Int. J. Wildland Fire* **2020**, *29*, 961. [CrossRef]
4. Butry, D.T.; Mercer, E.D.; Prestemon, J.P.; Pye, J.M.; Holmes, T.P. What Is the Price of Catastrophic Wildfire? *J. For.* **2001**, *99*, 9–17. [CrossRef]
5. Secretariat of the Convention on Biological Diversity. *Impacts of Human-Caused Fires on Biodiversity and Ecosystem Functioning, and Their Causes in Tropical, Temperate, and Boreal Forest Biomes*; Secretariat of the Convention on Biological Diversity: Montreal, QC, Canada, 2001.
6. Strid, A. The Greek Mountain Flora, with Special Reference to the Central European Element. *Bocconea* **1995**, *5*, 99–112.
7. Tan, K.; Stevanović, V.; Strid, A. Distribution and Centres of Diversity for Endemic Geophytic Monocots in the Balkans. *Bocconea* **2007**, *21*, 139–146.
8. Blandford, D. "Burn Baby Burn"—Controlling the Risk of Wildfires in Greece. In Proceedings of the 93rd Annual Conference, Warwick University, Coventry, UK, 15–17 April 2019.
9. Copernicus Emergency Management Service. Available online: https://Emergency.Copernicus.Eu/Mapping/List-of-Components/EMSR527 (accessed on 10 December 2021).
10. Field, C.B.; Barros, V.; Stocker, T.F.; Dahe, Q. *Managing the Risks of Extreme Events and Disasters to Advance Climate Change Adaptation: Special Report of the Intergovernmental Panel on Climate Change*; Field, C.B., Barros, V., Stocker, T.F., Dahe, Q., Eds.; Cambridge University Press: Cambridge, UK, 2012; ISBN 978-1-139-17724-5.
11. San-Miguel-Ayanz, J.; Durrant, T.; Boca, R.; Libertà, G.; Branco, A.; de Rigo, D.; Ferrari, D.; Maianti, P.; Artes, T.; Costa, H.; et al. *Forest Fires in Europe, Middle East and North Africa 2017*; Publications Officer of the European Union: Luxembourg, 2018; ISBN 978-92-79-92831-4.
12. Ganteaume, A.; Syphard, A.D. Ignition Sources. In *Encyclopedia of Wildfires and Wildland-Urban Interface (WUI) Fires*; Manzello, S.L., Ed.; Springer International Publishing: Cham, Switzerland, 2018; pp. 1–17. ISBN 978-3-319-51727-8.
13. Ganteaume, A.; Camia, A.; Jappiot, M.; San-Miguel-Ayanz, J.; Long-Fournel, M.; Lampin, C. A Review of the Main Driving Factors of Forest Fire Ignition Over Europe. *Environ. Manag.* **2013**, *51*, 651–662. [CrossRef]
14. Nuthammachot, N.; Stratoulias, D. Multi-Criteria Decision Analysis for Forest Fire Risk Assessment by Coupling AHP and GIS: Method and Case Study. *Environ. Dev. Sustain.* **2021**, *23*, 17443–17458. [CrossRef]
15. Dillon, G.K.; Holden, Z.A.; Morgan, P.; Crimmins, M.A.; Heyerdahl, E.K.; Luce, C.H. Both Topography and Climate Affected Forest and Woodland Burn Severity in Two Regions of the Western US, 1984 to 2006. *Ecosphere* **2011**, *2*, art130. [CrossRef]
16. Estes, B.L.; Knapp, E.E.; Skinner, C.N.; Miller, J.D.; Preisler, H.K. Factors Influencing Fire Severity under Moderate Burning Conditions in the Klamath Mountains, Northern California, USA. *Ecosphere* **2017**, *8*, e01794. [CrossRef]
17. Fang, L.; Yang, J.; Zu, J.; Li, G.; Zhang, J. Quantifying Influences and Relative Importance of Fire Weather, Topography, and Vegetation on Fire Size and Fire Severity in a Chinese Boreal Forest Landscape. *For. Ecol. Manag.* **2015**, *356*, 2–12. [CrossRef]
18. Birch, D.S.; Morgan, P.; Kolden, C.A.; Abatzoglou, J.T.; Dillon, G.K.; Hudak, A.T.; Smith, A.M.S. Vegetation, Topography and Daily Weather Influenced Burn Severity in Central Idaho and Western Montana Forests. *Ecosphere* **2015**, *6*, art17. [CrossRef]
19. Erten, E.; Kurgun, V.; Musaoglu, N. *Forest Fire Risk Zone Mapping from Satellite Imagery and GIS: A Case Study*; ISPRS: Bethesda, MD, USA, 2002.
20. Vasilakos, C.; Kalabokidis, K.; Hatzopoulos, J.; Matsinos, I. Identifying Wildland Fire Ignition Factors through Sensitivity Analysis of a Neural Network. *Nat. Hazards* **2009**, *50*, 125–143. [CrossRef]
21. Kopecký, M.; Macek, M.; Wild, J. Topographic Wetness Index Calculation Guidelines Based on Measured Soil Moisture and Plant Species Composition. *Sci. Total Environ.* **2021**, *757*, 143785. [CrossRef] [PubMed]
22. Zhao, L.; Yebra, M.; van Dijk, A.I.J.M.; Cary, G.J.; Matthews, S.; Sheridan, G. The Influence of Soil Moisture on Surface and Sub-Surface Litter Fuel Moisture Simulation at Five Australian Sites. *Agric. For. Meteorol.* **2021**, *298–299*, 108282. [CrossRef]
23. Pourtaghi, Z.S.; Pourghasemi, H.R.; Aretano, R.; Semeraro, T. Investigation of General Indicators Influencing on Forest Fire and Its Susceptibility Modeling Using Different Data Mining Techniques. *Ecol. Indic.* **2016**, *64*, 72–84. [CrossRef]
24. Adaktylou, N.; Stratoulias, D.; Landenberger, R. Wildfire Risk Assessment Based on Geospatial Open Data: Application on Chios, Greece. *IJGI* **2020**, *9*, 516. [CrossRef]
25. Zhao, P.; Zhang, F.; Lin, H.; Xu, S. GIS-Based Forest Fire Risk Model: A Case Study in Laoshan National Forest Park, Nanjing. *Remote Sens.* **2021**, *13*, 3704. [CrossRef]
26. Gai, C.; Weng, W.; Yuan, H. GIS-Based Forest Fire Risk Assessment and Mapping. In Proceedings of the 2011 Fourth International Joint Conference on Computational Sciences and Optimization, Kunming, China, 15–19 April 2011; IEEE: Manhattan, NY, USA, 2011; pp. 1240–1244.
27. Parajuli, A.; Gautam, A.P.; Sharma, S.P.; Bhujel, K.B.; Sharma, G.; Thapa, P.B.; Bist, B.S.; Poudel, S. Forest Fire Risk Mapping Using GIS and Remote Sensing in Two Major Landscapes of Nepal. *Geomat. Nat. Hazards Risk* **2020**, *11*, 2569–2586. [CrossRef]
28. Tomar, J.S.; Kranjčić, N.; Đurin, B.; Kanga, S.; Singh, S.K. Forest Fire Hazards Vulnerability and Risk Assessment in Sirmaur District Forest of Himachal Pradesh (India): A Geospatial Approach. *IJGI* **2021**, *10*, 447. [CrossRef]

29. Mu, E.; Pereyra-Rojas, M. Understanding the Analytic Hierarchy Process. In *Practical Decision Making*; Springer International Publishing: Cham, Switzerland, 2017; pp. 7–22. ISBN 978-3-319-33860-6.
30. Ajin, R.S.; Ciobotaru, A.-M.; Vinod, P.G.; Jacob, M.K. Forest and Wildland Fire Risk Assessment Using Geospatial Techniques: A Case Study of Nemmara Forest Division, Kerala, India. *J. Wetl. Biodivers.* **2015**, *5*, 29–37.
31. Nezhad, M.M.; Heydari, A.; Fusilli, L.; Laneve, G. Land Cover Classification by Using Sentinel-2 Images: A Case Study in the City of Rome. In Proceedings of the 4th World Congress on Civil, Structural, and Environmental Engineering (CSEE'19), Rome, Italy, 7–9 April 2019.
32. Cavur, M.; Duzgun, H.S.; Kemec, S.; Demirkan, D.C. Land Use and Land Cover Classification of Sentinel 2-A: St Petersburg Case Study. *Int. Arch. Photogramm. Remote Sens. Spat. Inf. Sci.* **2019**, *42*, 13–16. [CrossRef]
33. Usman, B. Satellite Imagery Land Cover Classification Using K-Means Clustering Algorithm Computer Vision for Environmental Information Extraction. *Elixir Int. J. Comput. Sci. Eng.* **2013**, *63*, 18671–18675.
34. Norovsuren, B.; Tseveen, B.; Batomunkuev, V.; Renchin, T.; Natsagdorj, E.; Yangiv, A.; Mart, Z. Land Cover Classification Using Maximum Likelihood Method (2000 and 2019) at Khandgait Valley in Mongolia. *IOP Conf. Ser. Earth Environ. Sci.* **2019**, *381*, 012054. [CrossRef]
35. Shi, D.; Yang, X. Support Vector Machines for Land Cover Mapping from Remote Sensor Imagery. In *Monitoring and Modeling of Global Changes: A Geomatics Perspective*; Li, J., Yang, X., Eds.; Springer Remote Sensing/Photogrammetry; Springer: Dordrecht, The Netherlands, 2015; pp. 265–279. ISBN 978-94-017-9812-9.
36. Matzarakis, A. *The Climate of Evros*; University of Freiburg: Freiburg, Germany, 2006; ISBN 978-3-00-020071-7.
37. The Dadia-Lefkimi-Soufli Forest National Park. Available online: http://dadia-np.gr/?lang=en (accessed on 22 January 2022).
38. Copernicus Land Monitoring Service. Available online: https://land.copernicus.eu/imagery-in-situ/eu-dem/eu-dem-v1.1 (accessed on 5 October 2021).
39. Schindler, S.; Poirazidis, K.; Wrbka, T. Towards a Core Set of Landscape Metrics for Biodiversity Assessments: A Case Study from Dadia National Park, Greece. *Ecol. Indic.* **2008**, *8*, 502–514. [CrossRef]
40. Copernicus Open Access Hub. Available online: https://scihub.copernicus.eu/dhus/#/home (accessed on 10 December 2021).
41. GEODATA.Gov. Available online: http://geodata.gov.gr/ (accessed on 20 October 2021).
42. Open Street Map. Available online: http://download.geofabrik.de/europe/greece.html?fbclid=IwAR2V9xnHZnHf0dx3ztBfJA6vkLtMIVLf0GIz1LyAN9x9QVQyie4ntrE9RqU (accessed on 22 December 2021).
43. The Dadia-Lefkimi-Soufli Forest National Park, Press Release, 07 October 2020. Available online: https://dadia-np.gr/?p=6370 (accessed on 22 January 2022).
44. The Dadia-Lefkimi-Soufli Forest National Park, Press Release, 14 July 2021. Available online: https://dadia-np.gr/?p=6538 (accessed on 22 January 2022).
45. NASA-FIRMS. Available online: https://localhost.modaps.eosdis.nasa.gov/map/ (accessed on 16 January 2022).
46. Shima, L.J.; Anderson, R.R.; Carter, V.P. The Use of Aerial Color Infrared Photography in Mapping the Vegetation of a Freshwater Marsh. *Chesap. Sci.* **1976**, *17*, 74. [CrossRef]
47. Xanthopoulos, G.; Calfapietra, C.; Fernandes, P. Fire Hazard and Flammability of European Forest Types. In *Post-Fire Management and Restoration of Southern European Forests*; Moreira, F., Arianoutsou, M., Corona, P., De las Heras, J., Eds.; Springer: Dordrecht, The Netherlands, 2012; Volume 24, pp. 79–92. ISBN 978-94-007-2207-1.
48. Romero-Calcerrada, R.; Novillo, C.J.; Millington, J.D.A.; Gomez-Jimenez, I. GIS Analysis of Spatial Patterns of Human-Caused Wildfire Ignition Risk in the SW of Madrid (Central Spain). *Landsc. Ecol.* **2008**, *23*, 341–354. [CrossRef]
49. Saaty, T.L. A Scaling Method for Priorities in Hierarchical Structures. *J. Math. Psychol.* **1977**, *15*, 234–281. [CrossRef]
50. Saaty, T.L. How to Make a Decision: The Analytic Hierarchy Process. *Eur. J. Oper. Res.* **1990**, *48*, 9–26. [CrossRef]
51. Samaniego, L.; Schulz, K. Supervised Classification of Agricultural Land Cover Using a Modified K-NN Technique (MNN) and Landsat Remote Sensing Imagery. *Remote Sens.* **2009**, *1*, 875–895. [CrossRef]
52. Daily Fire Risk Map, Ministry for Climate Crisis and Civil Protection. Available online: https://www.civilprotection.gr/el/archive/daily_map/2021 (accessed on 30 January 2022).
53. EFFIS—Current Situation. Available online: https://effis.jrc.ec.europa.eu/apps/effis_current_situation/ (accessed on 22 January 2022).
54. Sze, L.T.; Cheaw, W.G.; Ahmad, Z.A.; Ling, C.A.; Chet, K.V.; Lateh, H.; Bayuaji, L. High Resolution DEM Generation Using Small Drone for Interferometry SAR. In Proceedings of the 2015 International Conference on Space Science and Communication (IconSpace), Langkawi, Malaysia, 10–12 August 2015; pp. 366–369.

applied sciences

MDPI

Article

Visibility Assessment of New Photovoltaic Power Plants in Areas with Special Landscape Value

Enrique Zorzano-Alba, Luis Alfredo Fernandez-Jimenez *, Eduardo Garcia-Garrido, Pedro M. Lara-Santillan, Alberto Falces, Pedro J. Zorzano-Santamaria, Candido Capellan-Villacian and Montserrat Mendoza-Villena

Department of Electrical Engineering, University of La Rioja, 26004 Logroño, Spain; enrique.zorzano@unirioja.es (E.Z.-A.); eduardo.garcia@unirioja.es (E.G.-G.); pedro.lara@unirioja.es (P.M.L.-S.); alberto.falces@unirioja.es (A.F.); pedrojose.zorzano@unirioja.es (P.J.Z.-S.); candido.capellan@unirioja.es (C.C.-V.); montserrat.mendoza@unirioja.es (M.M.-V.)

* Correspondence: luisalfredo.fernandez@unirioja.es

Abstract: Power plants based on renewable sources offer environmental, technical and economic advantages. Of particular importance is the reduction in greenhouse gas emissions compared to conventional power plants. Despite the advantages, people are often opposed to the construction of these facilities due to their high visual impact, particularly if they are close to places with a great cultural and/or landscape value. This paper proposes a new methodology for identifying the most suitable geographical areas for the construction of new photovoltaic (PV) power plants in zones of special scenic or cultural interest, helping to keep the environment free from the visual intrusions caused by these facilities. From several repeated analyses, the degree of visibility of the new PV plant, the potential observation time of passing visitors, considering the route they follow and their speed, and the increase in visibility of the plants when seen totally or partially with the sky as background, are determined. The result obtained is a map showing the ranking of the geographical areas based on a variable calculated in such analyses: the Global Accumulated Perception Time (*GAPT*). The application of this methodology can help the different agents involved in the decision-making process for the installation of new PV plant by providing them with an objective visibility criterion.

Keywords: visual impact; landscape heritage; photovoltaic plants; geographical information systems; perception time; viewshed analysis

Citation: Zorzano-Alba, E.; Fernandez-Jimenez, L.A.; Garcia-Garrido, E.; Lara-Santillan, P.M.; Falces, A.; Zorzano-Santamaria, P.J.; Capellan-Villacian, C.; Mendoza-Villena, M. Visibility Assessment of New Photovoltaic Power Plants in Areas with Special Landscape Value. *Appl. Sci.* **2022**, *12*, 703. https://doi.org/10.3390/app12020703

Academic Editor: Yannis Maniatis

Received: 17 September 2021
Accepted: 10 January 2022
Published: 11 January 2022

1. Introduction

The integration of power plants based on renewable energies into the electrical grids has accelerated in recent years, primarily for environmental reasons, such as the effort to reduce greenhouse gas emissions from fossil fuels. Economic support from government authorities for these facilities, which are mainly wind and photovoltaic (PV) power generation plants, have led to their expansion throughout Europe [1]. For this expansion, finding the land available for building a new facility is the initial requirement. In the case of new wind farms or PV power plants, there are many factors involved in the decision to select the right location (wind or solar resources, distances to roads, distances to power lines, terrain orography, proximity to urban centres, visual impact, etc.) [2–5].

In areas where tourism is an important component of the local economy or in areas with a high landscape or cultural value, one of the most important factors in people accepting the installation of new power facilities is the visual impact. It is in the interest of local authorities and economic entities related to tourism that visitors or tourists perceive the landscape in its natural or original state [6] and as free from visual intrusions as possible. This dimension can cause projects to be suspended as a result of social rejection to the alteration or modification of the landscape [7]. In the field of power plants based on renewable energies, wind energy has the greatest negative impact on landscapes per unit of energy generation, followed by PV solar energy [8]. This has meant that the visual impact assessment of wind farms is more developed, although some of the proposed solutions can be applied to PV plants.

Cohen et al. in [9] seek to develop a conceptual definition of social acceptance, identifying and synthesizing the factors of discontent at play in the acceptance of wind farms, transmission lines and pumping groups for energy storage. In wind farm projects, the assessment of visual impact acquires great importance because of their lack of aesthetic integration into the landscape [10]: the authors assess the magnitude of the aesthetic impact on the landscape by means of a proposed indicator. Measurements of visibility, colour, fractality and continuity are taken from photographs and combined with each other. The results of the indicator are contrasted with the impact perceived by a sample of the local population. This perceived impact is found through surveys. In [11], the variables that can affect the visual impact are analysed, such as the visual magnitude and the overall colour difference, according to the author. From the analysis of images of the wind farm under study, parameters such as the difference in clarity, the difference in colour saturation and the difference in hue are obtained, which factor into the calculation of the overall difference in the colour represented by the visibility of the wind farm. The author analysed images of the wind farm at different times of the day and, using atmospheric visibility data, determined the temporal distribution of visual impacts in the area being studied. Other works use a method that quantifies the degree of visibility of an offshore wind farm from various observation points along the coast [12]. In this assessment of the degree of visibility, the authors introduce three indicators: the horizon occupation indicator (surface occupied by the wind farm on the horizon and defined by the area delimited by the convex envelope joining the turbine hubs), the distinguishable turbine indicator (defined as the relationship between the number of distinguishable turbines and the total number of turbines), and the aesthetic indicator (based on the alignment of the turbines).

PV power plants can bring about the transformation of a large area (land-use change, earthworks, vegetation removal), producing a significant alteration to the landscape. For assessing the aesthetic impact of PV plants, an indicator based on four parameters is proposed in [13], similar to the indicator presented in [10] for wind farms. In addition, in PV plants, there is a risk of glare from the reflection of sunlight on the surface of the PV modules, which makes them visible from great distances, producing a landscape alteration and negative visual impact on the environment [14].

Geographic information systems (GIS) are tools that make it possible to organize, analyse and model large amounts of spatial data, which facilitates the creation of maps for decision making in various types of projects. The scientific and technical literature includes many applications of GIS. Among other applications, they can be used to help in the assessment of renewable energy resources. For solar energy resources, Moser et al. [15] assess the PV potential in southern Tyrol, northern Italy, taking into account PV facilities on roofs and non-conventional surfaces. In a GIS, it is possible to perform a spatial analysis of the solar resource data together with other data from the area under survey, such as orography and distances to roads and power lines, or even take climatic data into consideration, which can help to locate the best sites for building PV plants [16]. A GIS system is also a good tool for visual impact assessment: it is possible to generate maps of the region under study that show those areas where the installation of PV power plants would have less impact, or areas where the installation of these facilities would not be suitable due to their high visual impact on the environment. Rodrigues et al. [17] address visual impact from spatial and perceptual points of view. From the spatial point of view, the result is a visibility map in which each cell (geographic position) has a Boolean value indicating whether an installation built in that position is seen from the observer's location, and from the perceptual point of view, a visual perception map is obtained in which each cell is assigned the affected angle of vision. The parameters required by the model are the dimensions of the facility (height and width), the visual threshold (maximum distance from which a facility can be recognized), and the height of the observer's position.

Manchado et al. [18] present a study that takes into account criteria of visibility and visual impact in the design stage of wind farms. The proposed methodology is based on two indices that describe the conditions of visual intrusion. The first index, called Magnitude of the Visual Effect (MVE), is the product of three visual indicators (visually affected area, visually affected population and visual exposure in linear sections), and the second one is an improved version of the Spanish Method index (SPM) proposed in [19]. Other authors take into account the visual capacity of the observer for visual impact assessment, using GIS tools to generate 3D maps [20]. For the assessment of the visual impact of wind turbines, other methods take into account the height of the visible part of the wind turbine and what percentage of it occupies the scene, for which GIS software together with 3D graphics generation software have been used [21]. Assessing the visual impact of PV plants using GIS-based tools helps to select geographic locations where the visual impact is lower. In [22], a relative visual impact index is defined, taking into account aspects such as the number of inhabitants in the surrounding area, the orography of the terrain and the height of the PV plants. From this index, visual impact maps are generated for two types of PV power facilities (with fixed panels or with trackers). Most of the GIS software have tools that make it possible to determine the visible geographic area (viewshed) from one or several observation points. Reference [23] proposes a methodology based on the fuzzy viewshed and the distance decay methods, which enables the calculation of the maximum number of hours in an average day in which a new PV plant can be seen by every possible observer. It takes into account all possible observers in motion (by roads) or in situ (urban centres), the orography of the terrain, the height of the observer and the size of the PV plant (height of panels, surface occupied). In urban areas, there is an increase in the use of solar PV technologies, which are mainly attached to building envelopes (roofs and facades). Florio et al. presents in [24] a methodology for assessing the visibility of building envelope surfaces exposed to solar radiation, which could host solar modules (thermal or photovoltaic) in urban areas and where public perception of this type of facility is not affected. The viewshed is determined on the cumulative viewing time from observation points, equidistantly arranged along urban roads, taking into account the height of the observer. The maximum visibility distance limit for calculations is 500 m, imposed by computational constraints.

As a consequence of population growth and industrial and socio-economic development, the need to build new infrastructure arises, which may have a direct impact on cultural and landscape heritage, the components of which may be seriously affected [25]. Therefore, the conflict between development policies and heritage conservation policies makes it necessary to propose and make use of tools that evaluate the effects produced by such infrastructures on cultural and landscape heritage, in order to maintain a balance between the two conflicting interests [26]. The methodology proposed in this work can help the agents involved to make decisions regarding the most suitable sites for the installation of new PV plants under an objective visibility criterion, keeping landscapes of special interest or unique cultural sites safe from the intrusion of these facilities, which could create a significant visual impact. The methodology helps to identify the most suitable sites for the construction of new PV plants using GIS tools on a digital surface model (DSM) and taking into account the possible observers who move through the paths of geographical areas that are especially protected by interest in their cultural and landscape value. These tools are integrated in the open-source software QGIS [27]. The proposed method considers the global visibility of future PV power plant elements, that is, the method evaluates what portion of the elements can be seen by the observers and also what part of the elements can be seen above the skyline. Facilities with the sky in the background have higher colour contrast and therefore greater visibility than facilities with the terrain in the background [28]. Variations in the visibility of PV facilities, caused by changes in colour contrasts during the day or by the weather, have not been considered in this work.

In the scientific literature, the issue of visual impact caused by PV facilities has generally been approached under subjective criteria, evaluating the visual perception people have of these types of facilities by means of surveys, photographs, 3D computer simulations, etc. To our knowledge, no published work proposes an objective method, based on visibility (total or partial), to identify locations where the construction of new PV facilities would produce a lower visual impact than elsewhere. Furthermore, there are not any published works related to site selection for new facilities in areas of special landscape and cultural protection, where the potential observers are visitors or tourists travelling in such environment. The methodology proposed in this paper aims to fill these gaps.

We define a variable called Global Accumulated Perception Time (*GAPT*), for the evaluation of the visibility of new PV facilities. This variable, which will be defined in detail in Section 2, is related to the cumulative total hours in a year in which a proposed PV plant can be seen by observers moving along the roads or paths in the area under study. The assessment of the *GAPT* variable in a geographical area makes it possible to obtain a set of GIS maps that help to visually identify the most suitable locations for the installation of new PV power plants in terms of their visibility.

In summary, the main objective of this work is the development of a new methodology based on GIS to determine, in an objective way, the places where future PV facilities will have a greater or lesser visibility for a set of potential observers in movement, and its application to places or geographical areas with special landscape or cultural protection.

The novel aspects covered in this work are:

- Assessment of the degree of visibility (total or partial) of the new PV plants;
- Assessment of the possible observation time of visitors or tourists, taking into account the route they follow and their speed;
- The proposed visibility enhancement factor for PV plants that may be fully or partially visible with the sky in the background.

The article is structured as follows: Section 2 presents the methodology used for the evaluation of the *GAPT* variable for all the areas surrounding the paths followed by the observers; Section 3 presents a case study with the application of the proposed methodology for the selection of suitable sites for the installation of two-axis PV trackers with an installed power capacity of 10 kW (the area under study corresponds to the area crossed by the Way of St. James in the region of La Rioja, Spain, declared a World Heritage Site by UNESCO) [29]; the results of the case study are shown and discussed in Section 4; finally, Section 5 presents the conclusions.

2. Methodology

The data on which a GIS runs are structured in layers containing information in vector or raster formats. Vector data represent geographic objects or entities such as points, nodes, lines or polygons. The values of the features of interest of such geographic objects are stored in the attribute table of the vector layer. Raster data are stored in an array of cells or pixels (each one representing an elementary geographic area) organized into rows and columns. The spatial resolution of a raster dataset determines the level of detail represented and depends on the size of the cell or geographic area it depicts. Both types of data are referenced to a geographic coordinate system.

In the proposed methodology, the main objective is to obtain a set of GIS maps that facilitate the identification of those places where the perception time of new PV power plants would be more reduced. The observers considered are visitors or tourists travelling on roads, paths or trails (we will refer to them as routes throughout the article) of the geographical area studied. GIS tools enable the creation of such maps working with data in raster format. These maps represent, for each geographic area or cell, the values of the variable called Global Accumulated Perception Time (*GAPT*). This variable corresponds to the accumulated value of the number of hours per year that a proposed PV plant can be seen by all possible observers in motion, considering all the observation points in the area under study. That is, the *GAPT* variable is the sum of the values of another variable, the Accumulated Perception Time (*APT*), for all the possible observation points. Considering a single observation point, the value of the variable *APT* in any cell represents the cumulative value of the possible hours per year in which the proposed PV plant, located in such a cell, can be seen by all observers passing through the observation point. It should be noted that, throughout this article, when we say that a PV plant is located in a particular cell, it really means that the PV plant is located in the elementary geographic area represented by that cell in a GIS.

2.1. Accumulated Perception Time

2.1.1. Required Data

The calculation of the hours per year, represented by the *APT* variable, takes into account increasing or decreasing factors. These factors include the distance between the observation point and the cell in which the proposed PV plant is located, the fraction of the PV plant that can be seen from the observation point, and the fraction that can be seen above the skyline. For the evaluation of the *APT* variable, it is necessary to consider aspects such as:

- Orography: Hills and depressions ensure that PV power plants remain hidden from the eyes of observers. In other areas, PV power plants can be fully or partially visible and some of their elements can be seen above the skyline, which increases their visibility. The orography is considered in a GIS using the digital elevation model (DEM) of the study area. The DEM is the digital representation of the elevation of the earth's surface with respect to a reference. Specifically, DEMs are a set that include digital terrain models (DTMs) and digital surface models (DSMs). DTMs represent the elevation of bare ground, while DSMs represent the elevation of the land surface, including obstacles not exclusively associated with terrain orography such as trees, vegetation, buildings, and other natural or artificial objects [30]. In this work, in order to consider visual obstacles on the ground, we have used a DSM of the analysed area. Different DSMs could also be used, as the density of vegetation can change over the seasons;
- Observation points: These represent the places where potential observers in motion can be located at a given time. These points are represented in a vector layer and have an associated attribute table containing the following data: geographic coordinates, height of the observer's eyes above the ground, height of the observed object above the ground, observation point elevation (z-coordinate), slope of the terrain in the direction of travel, and travel speed of the observers. From the speed value, it is possible to calculate the average observation time of the observers, as will be discussed in detail later in this section;
- Average annual number of observers travelling along the observation points of a given route;
- Colour contrast of the observed object with respect to the background. According to [31], objects with a higher colour contrast will have greater visibility than objects with a low contrast, therefore it is necessary to introduce a weighting factor as a function of this colour contrast of the facility;
- Distance between the observer and the observed object (proposed PV plant). According to [32], the visual acuity of the human being decreases with distance, therefore it is necessary to enter a weighting factor as a function of this distance.

Some works published in the scientific and technical literature consider this distance in models of viewshed analysis. Fisher in [33] proposes the use of fuzzy functions, the values of which decrease with the distance between observer and observed object. In the methodology used in this work, we have taken a similar approach, but we use a weighting function, shown in (1), with factors that have been adjusted taking into account both the height of the observed object and the loss of human visual acuity with distance, using logarithmic functions. This weighting function was used by Weigel in 2007 [34], improving the proposal of Paul in 2004 [35]. The weighting factors are limited by distance in three different perception zones as:

$$\begin{cases} w = 0.3 \ for \ d \leq 57.5 \\ w = \frac{-0.0638 \cdot \ln(d) + 0.59}{1.105} \ for \ 57.5 < d < 10,000 \\ w = 0 \ for \ d \geq 10,000 \end{cases} \quad (1)$$

where w represents the weighting factor, which can take values between 0 and 0.3, and d represents the distance, in meters, between the observer and the observed object.

In this work, we have assumed that the potential observers are moving at the average speed of a walking human being. However, the methodology is also applicable to other types of observers who travel by other means of locomotion (land vehicles, horseback, etc.). It is just a case of using the appropriate values of average speed and height of the observer's eyes above the ground when the observer is using these means of locomotion. The number of potential observers (*NOY*) corresponds to the annual number of walkers moving along a route (road, path or trail). *NOY* values can be obtained from local authorities, who can provide them for all of the routes included in the area studied.

2.1.2. Calculation Process

Suppose that along a given route, k, there are n nodes or arranged observation points. The value $APT_{i,k}$ corresponds to the sum of the values of the *APT* variable for all the n observation points of route k, considering a new PV plant placed in cell i. The stages that define the process of assessing the value of the $APT_{i,k}$ variable are described below:

1. Generation of the set of positions of observation points. The positions of the observation points are generated by taking equidistant nodes along the route at a distance equal to the size of the cell selected to represent the values of the *APT* variable. The nodes, stored in a vector layer, have an associated attribute table containing the information required for each point: the geographical coordinates, the height of the eyes of the observer above the ground, the height of the observed object above the ground, the elevation of the observation point (z-coordinate), the slope of the terrain in the direction of travel, the average speed of the observer as he/she moves from one node to the next one, and the average observation time in each node. These last values are calculated in the next two stages;
2. Evaluation of the walking speed of the observer. An observer walking along delimited routes over different types of terrain does not have a constant speed, as may occur when walking on flat terrain clear of obstacles. The observer's speed will generally be slower when walking over rough terrain with steep slopes. In order to take into account the difficulty of walking routes in rough terrain, the Modified Tobler's Function proposed by Márquez et al. in [36] has been used. It consists of an exponential function that provides the walking speed depending on the slope of the route section by which the potential observer is walking. This function is shown in (2), where ws_n is the walking speed (km/h) in node n and δ_n is the angle of the terrain slope, in degrees, in the usual direction of the hiker, in that node.

$$ws_n = 4.8 \cdot e^{-5.3 \ |0.7 \tan \delta_n + 0.03|} \quad (\text{km/h}) \quad (2)$$

To determine the angles of the terrain slopes δ_n in the direction of travel associated with each node, the following steps are performed:

- To each node the value that collects the DTM cell with the same coordinates is assigned. Let us call this value the z-coordinate of the node, which is stored in the attribute table;
- Knowing the difference between the values of the z-coordinate of nodes n and $n + 1$ and the distance d_n between them, it is easy to obtain the angle of the slope δ_n, when the observer moves from node n to $n + 1$. If δ_n has a positive value, it

is an upward slope, while if it is negative, it is a downward slope. The value obtained is stored in the register corresponding to node n in the attribute table;
- All the nodes of route k are analysed in the direction of travel, obtaining the values δ_n of each node;
- The value of the real distance (D_n) separating nodes n and n + 1 is determined. The value of the distance d_n corresponds to the projection on the horizontal plane of the real distance D_n;
- Then, applying the hiking function (2), the values of the observer's velocity ws_n at each node n are obtained and stored in the attribute table.

3. Calculation of the average observation time t_n in node n. This corresponds to the travel time used by the walker to travel from node n to node n + 1 along the route k. The time t_n is calculated as the distance between consecutive nodes D_n divided by the velocity of observer ws_n assigned to node n; $t_n = D_n/ws_n$. The value of t_n obtained for each node n is stored in the attribute table;
4. Determination of the distance between the new PV plant and the observation point. By using GIS tools, it is possible to determine the Euclidean distance $d_{i,n}$ between the area represented by cell i and the observation point represented by node n of route k. Subsequently, the weighting factor $w_{i,n}$ is obtained for each cell i, as a function of the distance $d_{i,n}$, using the expression previously shown in (1);
5. Determination of the visible height factor ($fhv_{i,n}$). This represents the visible part (in terms of height) of the future PV plant, placed in the cell i, when the observer is in the observation point represented by node n. Previously, the maximum height (H_{pv}) of the PV plant was divided into segments of equal length (h_{seg}). To determine $fhv_{i,n}$, for each node n, several repeated analyses are performed, following the steps outlined below:
   - To carry out a correct "visibility analysis" with a DSM, a new DSM must be generated in which the elevation of the observer in node n and of the PV plant placed in the cell i must correspond to the elevation values for bare ground in that position or cell, collected from the DTM;
   - Let h be the analysed height of the PV plant. In each analysis, h is increased by one segment of length h_{seg}, i.e., h ranges h_{seg} to the total height H_{pv} of the PV facility;
   - Using GIS tools, visibility analyses are performed to evaluate the visibility factor ($fv^h_{i,n}$) of the PV plant with a height h, placed in the cell i, when the observer is in node n. The result obtained for $fv^h_{i,n}$ will take the value 0, if from node n it is not possible to see the PV plant in cell i. On the contrary, $fv^h_{i,n}$ will take the value 1, if from node n it is possible to see the PV facility with a height of h meters. The results, after applying the analysis to all the cells in the study area, are collected in a binary raster, which only stores ones and zeros;
   - The value of h is increased by one segment and the visibility analysis is subsequently repeated from the same node n. The last analysis will be when h reaches the value of H_{pv}. As a result of each analysis for each value of h, a binary raster dataset of $fv^h_{i,n}$ is obtained;
   - The values of $fv^h_{i,n}$ obtained for each value of h are then summed. The result corresponds to the portion in meters of the PV plant height placed in the area represented by cell i that can be seen by the observer in node n. Finally, it is multiplied by the term h_{seg}/H_{pv}, as shown in expression (3), obtaining the visible height factor $fhv_{i,n}$, which represents the value per unit of the height H_{pv} seen from the observation point n.

$$fhv_{i,n} = \frac{h_{seg}}{H_{pv}} \sum_{h=h_{seg}}^{H_{pv}} fv^h_{i,n} \quad for \quad H_{pv} > 0 \tag{3}$$

6. Evaluation of the skyline index. Generally, facilities above the skyline (with the sky in the background) will have a higher colour contrast and, therefore, higher visibility than facilities below the skyline (with the terrain in the background) [37]. In order to take this aspect into account, a skyline index (*Isk*) is calculated. Considering a PV facility in cell i and an observer in node n, with the $Isk_{i,n}$ index it is possible to give more weight to the visibility of the facility which is seen partially or completely above the horizon. Therefore, it is necessary to determine what fraction, in terms of height, of the PV facility is visible from node n above the horizon line. This index is calculated using the expression shown in (4), where only positive values are considered:

$$\begin{cases} Isk_{i,n} = \frac{d_{i,n}\left(\tan\theta f_{i,n} - \tan\theta h_{i,n}\right)}{H_{pv}} \quad for \quad \theta f_{i,n} > \theta h_{i,n} \\ Isk_{i,n} = 0 \quad for \quad \theta f_{i,n} \le \theta h_{i,n} \end{cases} \tag{4}$$

where $\theta f_{i,n}$ is the angle of elevation of the line of sight [38] between the observation point represented by node n and the maximum height H_{pv} of the PV facility placed in the cell i, and $\theta h_{i,n}$ corresponds to the angle of elevation of the line of sight connecting the position of the observer's eyes in node n to the global horizon beyond cell i, as shown in Figure 1.

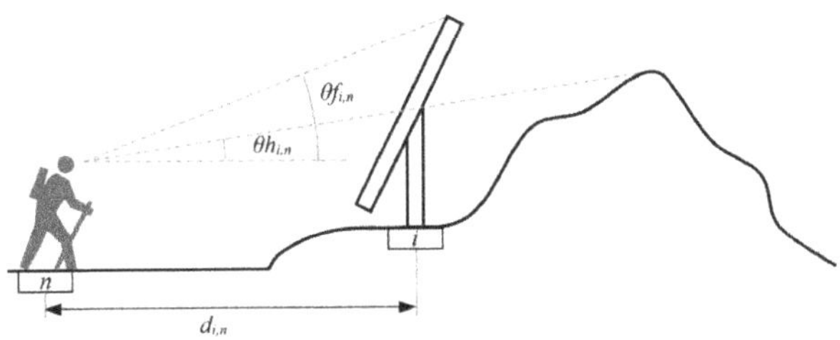

Figure 1. Lines of sight connecting the observer, in node n, to the PV facility and to the horizon.

The $Isk_{i,n}$ index can take values between 0 and 1, and represents the value per unit of the height H_{pv} above the skyline, as seen by an observer located at node n. When the whole facility (in terms of height) is below the skyline $Isk_{i,n}$ takes a value of 0, on the other hand, when the entire facility is above the skyline $Isk_{i,n}$ takes a value of 1. To evaluate $Isk_{i,n}$ it is necessary to determine the elevation angles $\theta h_{i,n}$ and $\theta f_{i,n}$ beforehand, as follows:

- First, we assume that observers can look in any direction, so it is necessary to determine the horizon line around each observation point or node n, whose geographical coordinates are known. That is, the horizon height $\theta h_{i,n}$ must be evaluated for observer azimuth values from 0 to 360 sexagesimal degrees. The observer's azimuth refers to the angle, measured on the horizontal plane, formed by the direction in which the observer is looking with respect to a reference direction. In our work, the azimuth value was 0 degrees when the observer was facing east and 90 degrees when facing north. After applying GIS tools, the result obtained, for each node n, is a raster dataset in which each cell i contains the value of the elevation angle $\theta h_{i,n}$ of the line of sight connecting the observer's eyes in node n to the global horizon beyond the area represented by the cell i. Consequently, all cells with the same azimuth value, will also take the same value of $\theta h_{i,n}$;
- To obtain the elevation angle $\theta f_{i,n}$, visibility analysis GIS tools are applied at each node n. As result, a raster dataset is obtained in which each cell i contains the value of the elevation angle $\theta f_{i,n}$ of the line of sight connecting the observer's eyes in node n with the highest part of the possible PV facility placed in the cell i;
- Subsequently, by means of raster data layer processing techniques (map algebra) the $Isk_{i,n}$ index is determined using the expression (4).

For a single observer in node n, the value of $APT_{i,n}$ in cell i will take the value resulting from the product of the visibility height factor $fhv_{i,n}$, the weighting factor $w_{i,n}$ as a function of distance $d_{i,n}$, the average observation time t_n, and the term (1 + $Isk_{i,n}$). This last term has been introduced in order to give a higher degree of visibility to facilities that can be seen above the skyline. We have applied this methodology to routes that are mainly taken by walkers in periods of good weather with clear skies and high colour contrast (spring, summer and autumn seasons). Therefore, the visibility of the fraction of the PV plant that can be seen above the horizon line has received twice as much weight as the rest of the facility. The $APT_{i,n}$ calculation process is shown in Figure 2.

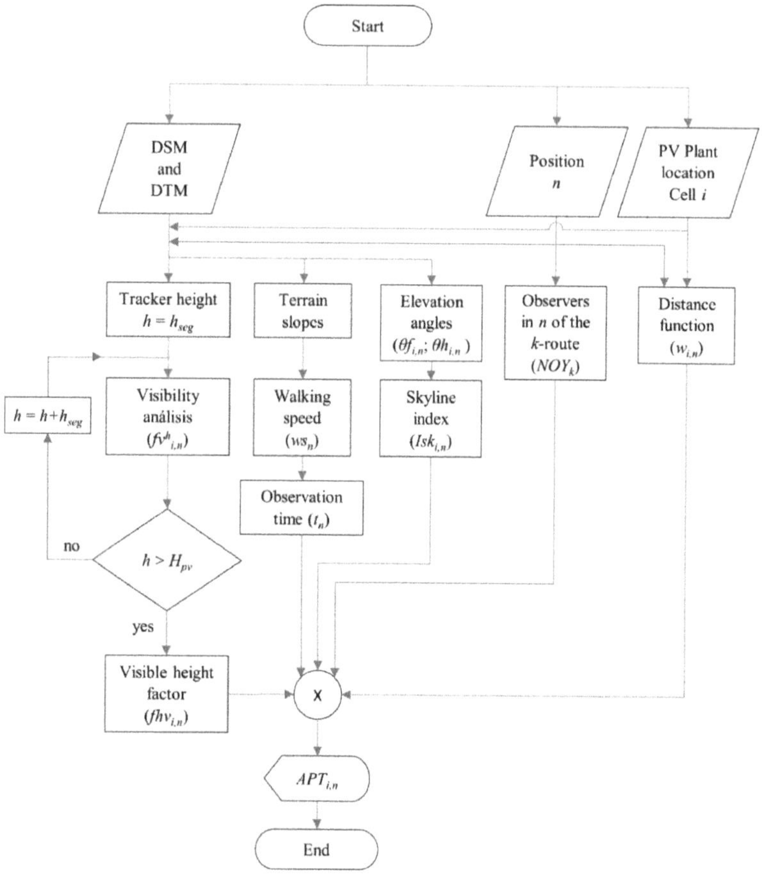

Figure 2. $APT_{i,n}$ calculation flow-chart for PV plant placed in the area represented by cell i, and observers in the position represented by node n.

Taking into account the average annual number of observers that can circulate through node n of route k (NOY_k), the cumulative value of $APT_{i,n}$ will be given by the expression (5).

$$APT_{i,n} = NOY_k \cdot fhv_{i,n} \cdot w_{i,n} \cdot t_n \cdot (1 + Isk_{i,n}) \qquad \text{(hours/year)} \tag{5}$$

2.2. Global Accumulated Perception Time

Finally, taking into account all the observation points or nodes n of route k and the number of observers NOY_k travelling through each route, the cumulative value of $APT_{i,k}$ in cell i, will be given by the expression (6),

$$APT_{i,k} = NOY_k \sum_{n=1}^{N_n} (fhv_{i,n} \cdot w_{i,n} \cdot t_n)(1 + Isk_{i,n}) \qquad \text{(hours/year)} \tag{6}$$

where N_n represents the total number of nodes or observation points on route k.

This process is repeated sequentially, analysing all possible routes (roads, path or trails used by the potential observers) in the area under study, so that for each cell i, the result of adding the $APT_{i,k}$ values of each route k is the global value of the accumulated perception time ($GAPT_i$), as shown in (7),

$$GAPT_i = \sum_{k=1}^{N_k} APT_{i,k} \quad \text{(hours/year)} \tag{7}$$

where N_k represents the total number of routes in the study area.

The $GAPT_i$ values in each of the cells in the studied area are collected in a raster dataset, which can be visually analysed in the form of a map. In this way, it is possible to easily visualize those areas with a lower *GAPT* value which are therefore candidates for the installation of PV plants under an objective visibility criterion. Note that the purpose of the proposed methodology is to be able to choose the place within the study area where a new PV plant with certain characteristics can be built with the least observability by visitors who move along the routes in that area.

This methodology can be applied to tourist areas or areas with unique qualities where, due to their location or their relationship with the landscape, it is necessary to determine those sites where the presence of new PV plants would be less harmful in terms of the visual impact. A similar case is its application to areas close to historical sites and landmarks, since PV facilities visible from these locations can have a negative impact on the perception of their cultural value [39]. The visual impact in places with a great cultural heritage has become a controversial topic [40] and, in general, the methodology proposed in this work can be used for evaluating the construction of new PV plants in locations that, due to their cultural heritage and their cultural, historical or landscape value, have to be preserved [41].

3. Case Study

We have applied the proposed methodology to the region of La Rioja in northern Spain. Specifically, the selected area of study corresponds to the area crossed by the Way of St. James. Figure 3 shows, in the bottom right corner, the region of La Rioja, and the studied area. The Way of Saint James is a world-renowned pilgrimage route that runs through northern Spain, from the western range of the Pyrenees to the city of Santiago de Compostela, in the northwest of the country. In 1993 it was declared a World Heritage Site by UNESCO, forming part of the cultural legacy of Europe, one of the most varied in the world [42].

In the studied area, the Way of St. James does not have a single route, rather it has different variants or branches that run through different towns and places with great cultural richness close to the main route and where it is possible to visit sanctuaries, temples and other monuments that are part of the cultural heritage of the region. The main route crossing La Rioja was established at the beginning of the 11th century. In addition to the historical and cultural interest of the Way, the different routes in La Rioja cross through vineyards that constitute an environment of special landscape beauty. The five routes of the Way in La Rioja are shown in Figure 3 and these are: main route, route of San Millán, route of Cirueña, route of Haro, and route of Briñas.

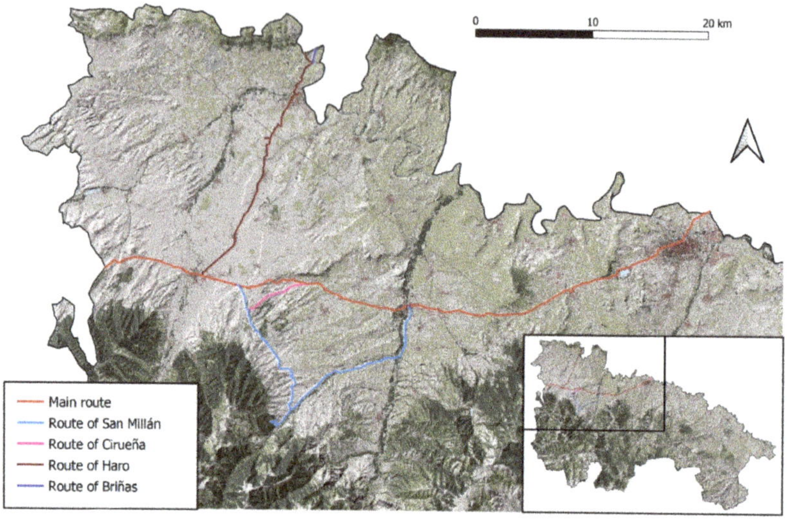

Figure 3. Way of Saint James through La Rioja (Spain).

In this work, the possible potential observers correspond to pilgrims walking along each of the five routes, while the inhabitants of the urban centres, included in the study area, have not been considered. Data related to pilgrimages have been taken from periodic statistical reports published by the Pilgrim's Office in the city of Santiago de Compostela. For this work, data referring to pilgrimages during the year 2019 have been used [43]. In that year, about 190,000 pilgrims walked these routes of the Way.

A PV system with a two-axis tracker has been selected for the construction of new PV plants in this case study. Each system can support up to 10 kW of power, depending on the type of PV module installed. The height of such a tracker is not constant throughout the day and its value varies depending on the elevation angle of the sun (angle between the sun's rays and the horizontal plane). The maximum height H_{pv} that can be reached by this facility is 6 m above the ground, mainly in the early and late hours of the day. This height was selected for this study because it is the most unfavourable value from the visibility criterion. The area occupied by a single tracker is approximately 625 m^2, including the terrain corresponding to transit areas.

In the GIS tool the main input data are the DSM and the DTM of the area of study. According to data from the National Geographic Institute of Spain [44], the DSM was obtained from flights made during the summer of 2016. This geographic data is composed of a set of square cells with an initial size of 2.5 × 2.5 m. Using resampling techniques, the original DSM was converted to a DSM with a cell size of 25 × 25 m. We used the nearest neighbour interpolation as the resampling method, which is one of the fastest interpolation methods. With this method, each cell or pixel of the resampled raster data acquires the same value as its nearest neighbour in the original raster. Originally, the DTM had a cell size of 25 × 25 m; therefore, it was not necessary to apply sampling techniques. The final cell size was chosen according to the size occupied by a two-axis solar tracker with a capacity of 10 kW. In other words, each solar tracker would occupy the geographic area represented by a single cell in GIS.

4. Results and Discussion

In order to evaluate the values of the *GAPT* variable in the region under study, the stages described above in Section 2 have been followed:

1. Using a GIS tool and considering the cell size of the DSM, equidistant nodes were generated every 25 m (d_n = 25 m) along each route of the Way of St. James. The points

corresponding to each of the five routes were collected in different vector data layers, in whose attribute tables, each node had associated data such as the average height above the ground of the observer's eyes (1.61 m in this study), and the maximum height H_{pv} above the ground of the PV facility (H_{pv} = 6 m);

2. GIS tools were used to obtain the value of the pilgrim's walking speed ws_n associated with each node n. This value was obtained applying expression (2). The value of walking speed was calculated for all the nodes of the five routes in the studied area;
3. The average observation time, t_n, was calculated for each node, dividing the real distance between consecutive nodes D_n by the value of the observer's walking speed ws_n at that node;
4. With the suitable GIS tool, a Euclidean distance raster map was generated for each node n, what allows to generate a new raster dataset containing the weight factor $w_{i,n}$;
5. Several repeated visibility analyses were performed to calculate the visible height factor $fhv_{i,n}$ of the future PV plant placed in the cell i, following the methodology described in Section 2. Note that the length of the segment used in this case study was 1 m (h_{seg} = 1 m) and, therefore, a set of six binary raster maps was obtained. Afterwards, using map algebra GIS tools, expression (3) was applied, obtaining a raster map with the values of the $fhv_{i,n}$ factor as a result, which represents the fraction of the PV facility located in the cell i, that can be seen from node n;
6. The skyline index $Isk_{i,n}$ was evaluated using the expression (4). For each node n, two raster maps were obtained, each one storing the values of $\theta h_{i,n}$ and $\theta f_{i,n}$, respectively, in cell i. Then, using map algebra techniques, the $Isk_{i,n}$ index was determined, which made it possible to give more weight to the visibility of future PV plants that could be seen above the skyline.

Next, taking into account the annual number of walkers traveling all the nodes of each of the five routes, NOY_k, the expression (6) was applied, obtaining a raster map in which each cell i represented the value of the accumulated perception time $APT_{i,k}$, achieved by the PV plant placed in that cell.

The volume of data in the raster format obtained in the calculation process was so high that R functions [45] were used to speed up this process and to automate access to the geo-algorithms integrated in QGIS from outside QGIS [46].

Finally, the expression (7) was applied to obtain the global value of accumulated perception time in each cell i ($GAPT_i$). Applying the described methodology to the selected study area, the $GAPT$ map shown in Figure 4 was obtained.

In the area of study, the urban centres are represented by yellow polygons and the different routes of the Way of St James are represented by red lines. The cells that show the DEM as background (green and grey cells at the top and bottom of Figure 4) represent the areas that have a zero $GAPT$ value and correspond to sites where the PV facilities are not visible from the routes, or to sites that are at a greater distance than the established limit of vision (10 km in our case study). These cells will be the best areas for the location of this type of solar facilities, from the point of view of their low observability from the routes of the Way of Saint James. The blue cells represent the lowest non-zero $GAPT$ values, while the red cells represent the zones with the highest $GAPT$ values and are therefore zones that should be avoided when it comes to the installation of PV plants. We can observe that the cells with higher $GAPT$ values are relatively near the main route (route 1). The main reason is that it corresponds to the route most used by pilgrims traveling to Santiago de Compostela. The routes that generally cross geographic areas with gentle depressions and elevations do not have too many natural obstacles and PV plants in such areas can reach significant values of cumulative perception time.

The lower right corner of Figure 4 shows a zoom of the area marked by a black rectangle. It is an area with gentle mountains and low vegetation, therefore, there is a greater number of cells with high $GAPT$ values. This is mainly due to the fact that a significant part of the facilities can be seen crossing the horizon line and having the sky as a background in these locations. Therefore, the observability of a PV plant such as the proposed one will be higher.

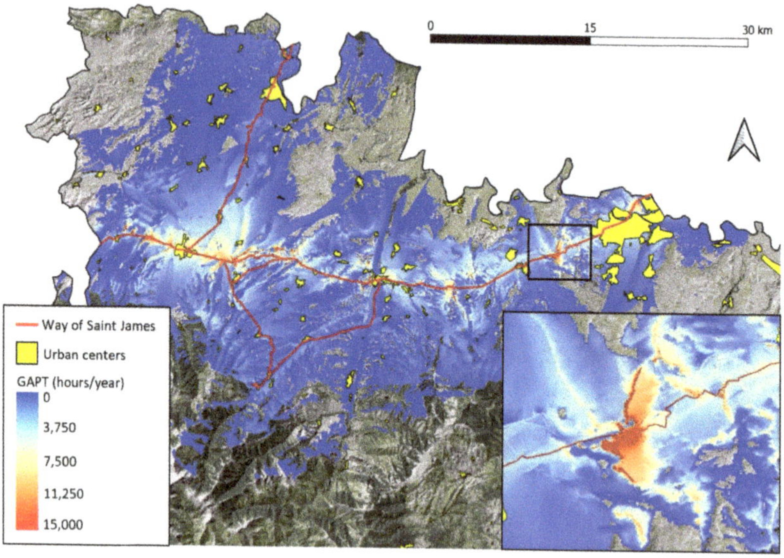

Figure 4. *GAPT* map, in hours/year, of the study area.

The results obtained provide accurate visibility maps with which it is easy to identify the locations where the installation of a future PV plant will have less visual impact for pilgrims walking along the Way of Saint James. The fundamental idea is to preserve the areas that surround and, above all, can be seen from the Way, so that they preserve its cultural legacy and previously held image.

In this case study, we have only analysed a 10 kW PV plant (each tracker occupies one cell of the map). Generally, the projected facilities will be much larger, and more cells will be required to cover the entire area occupied by the future facility. The solution to this problem is to select zones containing contiguous cells with low *GAPT* values.

5. Conclusions

The construction of new PV plants contributes to the reduction of CO_2 emissions but does not have the support of all the parties involved. In areas that have a high landscape value or in tourist areas, this type of facility can be opposed by the population due to its visual impact. This work has presented a methodology, based on GIS, to determine the places where future PV facilities will have a greater or lesser visibility for a set of potential observers in movement, and its application to geographical areas with special landscape or cultural protection. The methodology is based on an objective measure: the accumulated perception time for all the potential observers.

The perception that observers have of this type of facility is estimated by means of a variable called Accumulated Perception Time, $APT_{i,k}$. This variable represents the cumulative value of the possible hours per year in which the proposed PV plant, located in cell i, can be seen by observers travelling along all the observation points of route k. In order to make a more appropriate assessment of observability, two factors have been included in the calculation of $APT_{i,k}$: the factor $fhv_{i,n}$, which considers the visible part of the PV facility placed in the cell i, from observation point n, and the skyline index, $Isk_{i,n}$, which takes into account the part of the facility that can be see above the horizon. This makes it possible to present a greater degree of visibility to facilities that can be seen partially or totally above the horizon.

By calculating the values obtained for the $APT_{i,k}$ variables of each route k in the area studied, the $GAPT_i$ variable is evaluated. This variable represents the global cumulative value of hours per year in which the future PV plant, placed in the cell i, can be seen by all observers travelling along all routes k in the area under study. It is possible to generate maps that graphically represent the value of the *GAPT* variable, by means of the use of GIS tools. With these maps, the agents involved in the construction of PV plants can easily identify those locations in the study area with lower *GAPT* values and which will therefore be more suitable for the installation of PV plants, based objectively on their observability.

As a case study, we have applied this methodology to one of the most famous pilgrimage routes. Specifically, the study area is the section of the Way of Saint James that crosses the region of La Rioja in northern Spain. The observers that have been considered correspond to the pilgrims that walk along all routes of the Way in this region. The result is a raster dataset that represents the value of the *GAPT* variable, in which it is possible to identify those places which have lower values and which, therefore, will be more suitable for the installation of PV plants based on their low observability from the Way of Saint James.

Some of the limitations of the proposed methodology, that will be addressed in future works, focus on the following issues:

- A high computational effort is required to achieve detailed results. Each visibility analysis for different heights of the PV facility produces a raster map.
- All the nodes of the routes have been considered as waypoints for the observers. In a more detailed analysis, a special treatment could be considered for some of these nodes corresponding to places on the Way of St. James where pilgrims usually stop on their way: viewpoints, fountains or rest areas. The $APT_{i,n}$ value for these nodes should be proportional to the average stopping time of pilgrims at such locations.
- The visual impact caused by glares from PV panels has not been taken into account.

The methodology described here is easily applicable to different routes, observers with different characteristics, and photovoltaic facilities with different dimensions, etc. In such cases, it would suffice to introduce the appropriate values of the factors that intervene in the calculation of the *GAPT* variable. The results obtained are easily interpretable and can be used by decision-makers (investors and local authorities) to plan new PV plants in areas with special cultural, historical or landscape value. The methodology can also be applied to other types of facilities (wind turbines, electricity pylons, communications antennas and buildings), taking appropriate values for the cell size and the maximum height of the projected facility.

Author Contributions: Conceptualization, L.A.F.-J. and E.Z.-A.; methodology, L.A.F.-J. and E.Z.-A.; software, E.Z.-A., P.J.Z.-S., M.M.-V., E.G.-G. and C.C.-V.; validation, P.J.Z.-S., M.M.-V., P.M.L.-S. and C.C.-V.; formal analysis, E.Z.-A., M.M.-V. and A.F.; investigation, P.J.Z.-S., P.M.L.-S., E.G.-G. and C.C.-V.; resources, P.M.L.-S., E.G.-G., A.F. and C.C.-V.; data curation, P.J.Z.-S., M.M.-V., P.M.L.-S., A.F. and C.C.-V.; writing—original draft preparation, E.Z.-A.; writing—review and editing, L.A.F.-J. and E.Z.-A.; visualization, L.A.F.-J., P.J.Z.-S., M.M.-V., P.M.L.-S., E.G.-G. and A.F.; supervision, L.A.F.-J.; project administration, L.A.F.-J.; funding acquisition, L.A.F.-J. All authors have read and agreed to the published version of the manuscript.

Funding: This research was funded by University of La Rioja and Banco Santander (REGI2020-28).

Institutional Review Board Statement: Not applicable.

Informed Consent Statement: Not applicable.

Data Availability Statement: Not applicable.

Conflicts of Interest: The authors declare no conflict of interest.

Abbreviations

DEM	Digital Elevation Model
DSM	Digital Surface Model
DTM	Digital Terrain Model
GIS	Geographic Information System
PV	Photovoltaic
$APT_{i,n}$	Accumulated perception time in cell i, with the observers located at node n (h/y)
$APT_{i,k}$	Accumulated perception time in cell i by all observers moving along route k (h/y)
d_n	Distance in the horizontal plane between nodes n and $n + 1$ (meters)
D_n	True distance between nodes n and $n + 1$ (meters)
$d_{i,n}$	Euclidean distance between cell i and node n (meters)
$fhv_{i,n}$	Visible height factor of a PV plant placed in cell i and viewed from node n
$fv^h_{i,n}$	Visibility factor of a PV plant of height h, placed in cell i and viewed from node n
$GAPT_i$	Global Accumulated Perception Time in cell i
h	Analysed height of the PV plant (meters)
H_{pv}	Maximum height of the PV plant (meters)
h_{seg}	Segment length by which the height of the PV plant is increased in the visibility analysis
i	Geographic elemental area or cell with the possibility of housing a PV plant
$Isk_{i,n}$	Skyline index for a PV plant placed in cell i and viewed from node n
k	Any route considered in the study area
n	Observation point or node contained in route k
N_k	Total number of routes in the study area
N_n	Total number of nodes on route k
NOY_k	Annual number of potential observers moving along the route k
t_n	Average observation time when the observer moves from node n to node $n + 1$ (hours)
$w_{i,n}$	Weighting factor calculated as a function of distance $d_{i,n}$
ws_n	Walking speed of the observer at node n (km/h)
δ_n	Angle of the terrain slope in the direction of travel of the observer
$\theta f_{i,n}$	Elevation angle of the line of sight between node n and the top of the PV plant in cell i
$\theta h_{i,n}$	Elevation angle of the line of sight connecting the observer's eyes at n with the horizon, beyond cell i

References

1. Bachner, G.; Steininger, K.W.; Williges, K.; Tuerk, A. The economy-wide effects of large-scale renewable electricity expansion in Europe: The role of integration costs. *Renew. Energy* **2019**, *134*, 1369–1380. [CrossRef]
2. Al Garni, H.Z.; Awasthi, A. Solar PV power plant site selection using a GIS-AHP based approach with application in Saudi Arabia. *Appl. Energy* **2017**, *206*, 1225–1240. [CrossRef]
3. Carrión, J.A.; Estrella, A.E.; Dols, F.A.; Toro, M.Z.; Rodríguez, M.; Ridao, A.R. Environmental decision-support systems for evaluating the carrying capacity of land areas: Optimal site selection for grid-connected photovoltaic power plants. *Renew. Sustain. Energy Rev.* **2008**, *12*, 2358–2380. [CrossRef]
4. Xu, Y.; Li, Y.; Zheng, L.; Cui, L.; Li, S.; Li, W.; Cai, Y. Site selection of wind farms using GIS and multi-criteria decision making method in Wafangdian, China. *Energy* **2020**, *207*, 118222. [CrossRef]
5. Eroğlu, H. Multi-criteria decision analysis for wind power plant location selection based on fuzzy AHP and geographic information systems. *Environ. Dev. Sustain.* **2021**, *23*, 18278–18310. [CrossRef]
6. Terkenli, T.; Skowronek, E.; Georgoula, V. Landscape and Tourism: European Expert Views on an Intricate Relationship. *Land* **2021**, *10*, 327. [CrossRef]
7. Mouflis, G.D.; Gitas, I.Z.; Iliadou, S.; Mitri, G.H. Assessment of the visual impact of marble quarry expansion (1984–2000) on the landscape of Thasos island, NE Greece. *Landsc. Urban Plan.* **2008**, *86*, 92–102. [CrossRef]
8. Ioannidis, R.; Koutsoyiannis, D. A review of land use, visibility and public perception of renewable energy in the context of landscape impact. *Appl. Energy* **2020**, *276*, 115367. [CrossRef]
9. Cohen, J.J.; Reichl, J.; Schmidthaler, M. Re-focussing research efforts on the public acceptance of energy infrastructure: A critical review. *Energy* **2014**, *76*, 4–9. [CrossRef]
10. Sibille, A.D.C.T.; Cloquell-Ballester, V.-A.; Cloquell-Ballester, V.-A.; Darton, R. Development and validation of a multicriteria indicator for the assessment of objective aesthetic impact of wind farms. *Renew. Sustain. Energy Rev.* **2009**, *13*, 40–66. [CrossRef]
11. Bishop, I.D. The implications for visual simulation and analysis of temporal variation in the visibility of wind turbines. *Landsc. Urban Plan.* **2019**, *184*, 59–68. [CrossRef]

12. Maslov, N.; Claramunt, C.; Wang, T.; Tang, T. Method to estimate the visual impact of an offshore wind farm. *Appl. Energy* **2017**, *204*, 1422–1430. [CrossRef]
13. Torres-Sibille, A.d.C.; Cloquell-Ballester, V.A.; Cloquell-Ballester, V.A.; Artacho Ramírez, M.Á. Aesthetic impact assessment of solar power plants: An objective and a subjective approach. *Renew. Sustain. Energy Rev.* **2009**, *13*, 986–999. [CrossRef]
14. Chiabrando, R.; Fabrizio, E.; Garnero, G. The territorial and landscape impacts of photovoltaic systems: Definition of impacts and assessment of the glare risk. *Renew. Sustain. Energy Rev.* **2009**, *13*, 2441–2451. [CrossRef]
15. Moser, D.; Vettorato, D.; Vaccaro, R.; Del Buono, M.; Sparber, W. The PV Potential of South Tyrol: An Intelligent Use of Space. *Energy Procedia* **2014**, *57*, 1392–1400. [CrossRef]
16. Munkhbat, U.; Choi, Y. GIS-Based Site Suitability Analysis for Solar Power Systems in Mongolia. *Appl. Sci.* **2021**, *11*, 3748. [CrossRef]
17. Rodrigues, M.; Montañés, C.; Fueyo, N. A method for the assessment of the visual impact caused by the large-scale deployment of renewable-energy facilities. *Environ. Impact Assess. Rev.* **2010**, *30*, 240–246. [CrossRef]
18. Manchado, C.; Gomez-Jauregui, V.; Lizcano, P.E.; Iglesias, A.; Galvez, A.; Otero, C. Wind farm repowering guided by visual impact criteria. *Renew. Energy* **2019**, *135*, 197–207. [CrossRef]
19. Hurtado, J.; Fernández, J.; Parrondo, J.L.; Blanco, E. Spanish method of visual impact evaluation in wind farms. *Renew. Sustain. Energy Rev.* **2004**, *8*, 483–491. [CrossRef]
20. Molina-Ruiz, J.; Sánchez, M.J.M.; Pérez-Sirvent, C.; Serrano, M.L.T.; Lorenzo, M.L.G. Developing and applying a GIS-assisted approach to evaluate visual impact in wind farms. *Renew. Energy* **2011**, *36*, 1125–1132. [CrossRef]
21. Wróżyński, R.; Sojka, M.; Pyszny, K. The application of GIS and 3D graphic software to visual impact assessment of wind turbines. *Renew. Energy* **2016**, *96*, 625–635. [CrossRef]
22. Garcia-Garrido, E.; Lara-Santillan, P.; Zorzano-Alba, E.; Mendoza-Villena, M.; Zorzano-Santamaria, P.; Fernandez-Jimenez, L.A.; Falces, A. Visual impact assessment for small and medium size PV plants. In Proceedings of the 12th WSEAS International Conference on Electric Power Systems, High Voltages, Electric Machines, Prague, Czech Republic, 24–26 September 2012; pp. 57–61.
23. Fernandez-Jimenez, L.A.; Mendoza-Villena, M.; Zorzano-Santamaria, P.; Garcia-Garrido, E.; Lara-Santillan, P.; Zorzano-Alba, E.; Falces, A. Site selection for new PV power plants based on their observability. *Renew. Energy* **2015**, *78*, 7–15. [CrossRef]
24. Florio, P.; Probst, M.C.M.; Schüler, A.; Roecker, C.; Scartezzini, J.-L. Assessing visibility in multi-scale urban planning: A contribution to a method enhancing social acceptability of solar energy in cities. *Sol. Energy* **2018**, *173*, 97–109. [CrossRef]
25. Pham, V.-M.; Van Nghiem, S.; Van Pham, C.; Luu, M.P.T.; Bui, Q.-T. Urbanization impact on landscape patterns in cultural heritage preservation sites: A case study of the complex of Huế Monuments, Vietnam. *Landsc. Ecol.* **2021**, *36*, 1–26. [CrossRef]
26. Ashrafi, B.; Kloos, M.; Neugebauer, C. Heritage Impact Assessment, beyond an Assessment Tool: A comparative analysis of urban development impact on visual integrity in four UNESCO World Heritage Properties. *J. Cult. Herit.* **2021**, *47*, 199–207. [CrossRef]
27. QGIS. A Free and Open Source Geographic Information System. Available online: https://www.qgis.org/en/site/ (accessed on 28 January 2021).
28. Fisher, P.F. Extending the applicability of viewsheds in landscape planning. *Photogramm. Eng. Remote Sens.* **1996**, *62*, 1297–1302.
29. UNESCO. Routes of Santiago de Compostela: Camino Francés and Routes of Northern Spain. Available online: https://whc.unesco.org/en/list/669 (accessed on 10 September 2021).
30. Polidori, L.; El Hage, M. Digital Elevation Model Quality Assessment Methods: A Critical Review. *Remote. Sens.* **2020**, *12*, 3522. [CrossRef]
31. Bishop, I. Testing perceived landscape colour difference using the Internet. *Landsc. Urban Plan.* **1997**, *37*, 187–196. [CrossRef]
32. Ogburn, D.E. Assessing the level of visibility of cultural objects in past landscapes. *J. Archaeol. Sci.* **2006**, *33*, 405–413. [CrossRef]
33. Fisher, P.F. Probable and fuzzy models of the viewshed operation. In *Innovations in GIS*; CRC Press: Boca Raton, FL, USA, 1994; pp. 161–175.
34. Weigel, J. Kompensationsflächenberechnung für Freileitungen. Available online: https://www.yumpu.com/de/document/read/6771445/kompensationsflachenberechnung-fur-freileitungen (accessed on 31 August 2021).
35. Paul, H.-U.; Uther, D.; Neuhoff, M.; Winkler-Hartenstein, K.; Schmidtkunz, H.; Grossnick, J. GIS-gestütztes Verfahren zur Bewertung visueller Eingriffe durch Hochspannungsfreileitungen. *Nat. Und Landschaftsplan. Z. Für Angew. Okol.* **2004**, *36*, 139–144.
36. Pérez, J.M.; Vallejo, I.; Álvarez-Francoso, J.I. Estimated travel time for walking trails in natural areas. *Geogr. Tidsskr.-Danish J. Geogr.* **2017**, *117*, 53–62. [CrossRef]
37. Fontani, F. Application of the Fisher's "Horizon Viewshed" to a proposed power transmission line in Nozzano (Italy). *Trans. GIS* **2016**, *21*, 835–843. [CrossRef]
38. Caha, J.; Rášová, A. Line-of-Sight Derived Indices: Viewing Angle Difference to a Local Horizon and the Difference of Viewing Angle and the Slope of Line of Sight. *Lect. Notes Geoinf. Cartogr.* **2015**, *211*, 61–72. [CrossRef]
39. Serra, J.; Llinares, C.; Iñarra, S.; Torres, A.; Llopis, J. Improvement of the integration of visually impacting architectures in historical urban scene, an application of semantic differencial method. *Environ. Impact Assess. Rev.* **2019**, *81*, 106353. [CrossRef]
40. Jerpåsen, G.B.; Larsen, K.C. Visual impact of wind farms on cultural heritage: A Norwegian case study. *Environ. Impact Assess. Rev.* **2011**, *31*, 206–215. [CrossRef]

41. CETS 199—Council of Europe Framework Convention on the Value of Cultural Heritage for Society. Available online: https://rm.coe.int/1680083746 (accessed on 6 September 2021).
42. European Commission Directorate-General for Research and Innovation. *Preserving our Heritage, Improving our Environment. Volume I, 20 Years of EU Research into Cultural Heritage*; Chapuis, M., Ed.; Publications Office of the European Union: Luxembourg, 2009. Available online: https://data.europa.eu/doi/10.2777/17146 (accessed on 3 September 2021).
43. Informe Estadístico. 2019. Available online: http://oficinadelperegrino.com/wp-content/uploads/2016/02/peregrinaciones2019.pdf (accessed on 24 August 2021).
44. Instituto Geográfico Nacional. Available online: https://www.ign.es/web/ign/portal (accessed on 4 September 2021).
45. R Development Core Team. The R project for statistical computing, R version 3.5.2. 2018. Available online: https://www.r-project.org (accessed on 2 September 2021).
46. Lovelace, R.; Nowosad, J.; Muenchow, J. *Geocomputation with R*; Chapman and Hall/CRC: Boca Raton, FL, USA, 2019.

applied sciences

MDPI

Article

Application of Geoinformation Systems for Assessment of Effective Integration of Renewable Energy Technologies in the Energy Sector of Ukraine

Olga Ostapenko [1], Piotr Olczak [2], Viktor Koval [3,*], Larysa Hren [4], Dominika Matuszewska [5] and Olena Postupna [6]

1 Department of Heat Power Engineering, Vinnytsia National Technical University, 21021 Vinnytsia, Ukraine; ostapenko@vntu.edu.ua
2 Mineral and Energy Economy Research Institute, Polish Academy of Sciences, 31-261 Krakow, Poland; olczak@min-pan.krakow.pl
3 National Academy of Sciences of Ukraine, 01030 Kyiv, Ukraine
4 Kharkiv Polytechnic Institute, National Technical University, 61002 Kharkiv, Ukraine; Larysa.Hren@khpi.edu.ua
5 Department of Thermal and Fluid Flow Machines, Faculty of Energy and Fuels, AGH University of Science and Technology, 30 Mickiewicza Ave., 30-059 Krakow, Poland; dommat@agh.edu.pl
6 Training Research and Production Center, National University of Civil Defence of Ukraine, 61023 Kharkiv, Ukraine; nuczu@dsns.gov.ua
* Correspondence: koval@otei.odessa.ua or victor-koval@ukr.net

Abstract: The scientific novelty of the results presented in this article is to substantiate and expand the possibilities of using global and local geographic information systems (GIS) to assess the potential of renewable energy sources in Ukraine. GIS analysis focused on key resource parameters can help identify territories for development of renewable energy sources and assess of their possible technical potential, as well as the possibility of effective integration of technologies for the use of renewable energy sources in the energy sector of Ukraine. In this paper the possibilities for using geographic information systems to assess the potential of renewable energy sources in Ukraine are analyzed. The possibility of using the Global Atlas of the International Renewable Energy Agency (IRENA) to support planning of technologies for the use of energy from biomass is analyzed. The data can point to large-scale programs and applications in relation to key parameters (quality resources, transmission distance, population density, terrain and site protection), helping identify additional areas for development of renewable energy sources and give an approximate assessment of technical potential. It is determined that the software products of IRENA are able to support national and regional planning of renewable energy technologies, help establish the viability of future renewable energy facilities and help project developers identify and analyze promising facilities for the implementation of technologies using renewable energy. The application of geographic information systems of Ukrainian web resources ("UA MAP") for assessing the potential of renewable energy sources and energy efficiency in Ukraine is been analyzed. The scientific novelty of the results lies in applying global and local GIS for comprehensive assessment of the potential and effectiveness of the use of regional non-traditional and renewable energy resources on the territory of Ukraine. This makes it possible to assess the possibilities of generating additional electric and thermal power for the needs of the regions of Ukraine using non-traditional and renewable energy sources. A comprehensive methodology for the use of GIS is proposed for assessing the potential of non-traditional and renewable energy sources at the regional level in Ukraine, taking into account energy, environmental and socio-economic factors affecting the placement of non-traditional and renewable energy facilities.

Keywords: geoinformation systems; renewable energy sources; renewable energy technologies; Ukraine

Citation: Ostapenko, O.; Olczak, P.; Koval, V.; Hren, L.; Matuszewska, D.; Postupna, O. Application of Geoinformation Systems for Assessment of Effective Integration of Renewable Energy Technologies in the Energy Sector of Ukraine. *Appl. Sci.* **2022**, *12*, 592. https://doi.org/10.3390/app12020592

Academic Editor: Yannis Maniatis

Received: 3 December 2021
Accepted: 4 January 2022
Published: 7 January 2022

1. Introduction

Important conditions for achieving energy security and energy independence are the reduction of energy consumption and energy intensity of production, by increasing the energy efficiency of technological processes using non-traditional and renewable energy sources of natural and man-made origin [1]. In developing countries, it is necessary to introduce large-scale modern technologies for the use of renewable and non-traditional energy sources, the pace of implementation of which is not yet fully consistent with the European level [2–4]. However, in the foreseeable future we should expect the development of renewable energy in Ukraine.

In addition to the typical use of geographic information systems (GIS) in the areas of business and commercial activities and the use of natural resources, GIS is widely used in the public sector, transport and utilities. It is important for the state to have an adequate analysis of data in order to effectively identify the most in-demand and scarce resources in a timely manner.

GIS are systems for collecting, accumulating, processing and graphical visualization of spatial (geographical) data about GIS objects. The main areas of GIS use are urban and regional planning, economic development (GIS applications provide detailed analysis of social, economic and topographic features), emergencies and disaster management (assessment and monitoring of the environment, modeling of ecological disasters and analysis of their consequences, environmental planning), law enforcement, oil and gas industry, roads and transport, geodesy and logistics of infrastructure.

The purpose of GIS is determined by the tasks associated with decision-making in the field of environmental design/planning, rational use of natural resources, the adoption of sanitary protection measures in emergencies, and so on.

It should be noted that at present, extensive data sets have been formed on the basis of global and regional GIS systems, which can become the basis for research in the fields of non-traditional and renewable energy. In parallel with the problems associated with the verification of data from GIS systems and the analysis of the adequacy of methods for obtaining these information data, a number of problems should be noted that arise due to the complexity of the visual display of the required data in a form convenient for analysis. An important task in processing information data on non-traditional and renewable energy sources is their accumulation in suitable databases, as well as mapping the potential of non-traditional and renewable energy sources for various regions of Ukraine.

In addition to substantiated scientific and methodological significance, studies of the regional potential of non-traditional and renewable energy sources in Ukraine are of great practical relevance. The analyzed databases and GIS should become an influential tool for analyzing the effectiveness of the practical use of non-traditional and renewable energy sources in the regions of Ukraine, as well as for supporting the adoption of sound technical and managerial decisions on the use of potential regional non-traditional and renewable energy sources.

The relevance of the present research is determined by the need to create methods for using the resource potential of GIS in the field of non-traditional and renewable energy sources, taking into account the possible prerequisites and restrictions for their development at the regional level. This approach makes it possible to carry out zoning of territories in order to select the most promising sites for projects for the development of non-traditional and renewable energy in the regions of Ukraine.

The use of GIS permits prompt and detailed analysis of the studied indicators, to assess the possibilities and potential of regional non-traditional and renewable energy sources. In particular, GIS permits assessment of the possibilities of generating additional electricity and heat for the needs of regions, and the potential effectiveness of alternative non-traditional energy. The use of GIS to substantiate and develop technologies for the use of renewable energy sources will solve a number of energy, environmental and economic problems of the regions of Ukraine.

2. Methodology

In recent years, a number of scientific studies by various authors have been devoted to the application of geographic information systems to assess the potential of renewable energy sources [1–15]. As noted in a recent study [5], the relevance of this scientific topic is due to the fact that the use of renewable energy sources is assessed by the world community as one of the most promising ways to address growing energy supply and sustainable development needs [6].

In particular, the aim of these study [5] was to solve the current fundamental problem of environmentally safe use of a number of renewable energy sources in the Carpathian region of Ukraine, in line with the concept of sustainable development [7,8]. This was ensured by developing a scientific and methodological framework for strategic assessment of capacity and environmentally safe location of renewable energy sources, taking into account the sustainable development of the region [9,10], determination of optimal technologies for the use of renewable energy such as solar, wind and small hydropower, and stabilization and improvement of the environment based on the principles of sustainable balanced development of the region [10–12].

In [5], the scientific substantiation of technically achievable, economically expedient and ecologically safe potentials of renewable energy sources through creation of a complex of maps of GIS-potentials of renewable energy sources in the region was presented. A number of technical and economic advantages, technological and environmental priorities for the studied renewable energy sources were identified [8–11].

The first study [5] presented research on the development of a methodology for environmentally safe use of renewable energy sources in the Carpathian region of Ukraine, taking into account the concept of sustainable development. The scientific novelty of the results was to justify the expansion of the resource potential of renewable energy sources in the Carpathian region of Ukraine with the creation of a set of maps in the geographic information system "Map Information". For each of the types of renewable energy sources (solar, wind, small hydropower) in the study, a number of technical advantages have been identified, and technological and ecologically safe priorities for their use have been assessed. The article provided a detailed calculation of wind, solar and hydropower regional potentials of renewable energy sources for the Carpathian region of Ukraine. The spatial limitations and possibilities of introduction of the considered renewable energy sources in the context of sustainable development of the region of Ukraine were scientifically substantiated. A number of scenarios for the use of renewable energy in Ukraine were proposed.

GIS are used for graphical construction of maps and for obtaining information on individual facilities and spatial data on oblasts, such as the location of natural gas reserves, the density of transport communications or the distribution of per capita income in the state. The areas indicated on maps in many cases reflect the required information much more clearly than dozens of pages of reports with tables. The effective application of GIS is based on mathematical modeling in order to ensure the effectiveness of monitoring the effectiveness of nature management.

In recent years, there has been a steady growth in global investment in renewable energy. In 2015, the share of renewable energy sources in the structure of new installed energy capacity in the world reached a record level at 54%, which confirmed the long-term global trend in replacing traditional energy generation with renewable energy generation. In 2016, the share of renewable energy technologies in Europe accounted for 87% of all new installed capacity, which confirms the transformation of the European energy system in terms of replacing traditional renewable energy sources.

Our previous study [8] assessed the future trends in the reform of the energy sector of the world, the European Union and Ukraine up to 2050, using renewable energy sources and the concept of sustainable development. Our study [8] identified the benefits of using renewable energy sources and assessed the prospects for the use of innovative technologies based on renewable energy sources and the concept of sustainable development. A number of criteria for energy, economic and environmental efficiency of innovative technologies for the use of renewable energy sources were analyzed, to ensure comprehensive assessment of the effectiveness of energy- and resource-saving, environmentally-friendly and cost-effective innovative technologies in the field of sustainable development. This approach allowed us to rationally determine the prospects for the use of energy and resource-saving, environmentally-friendly and cost-effective innovative technologies for the use of renewable energy sources in line with the concept of sustainable development, in order to increase energy and economic efficiency and the level of environmental safety of the energy sector of Ukraine [8].

In our study, an energy model for the use of renewable energy sources based on global and local GIS data was created and investigated. Thus, in our study, data from different GIS sources were used in terms of time and are explored to analyze the possibilities of using alternative renewable energy sources in the regions. Statistical data from global and regional statistical resources also were used.

The general approach to modeling in our studies is based on the system approach and system analysis, which involves the selection of the main object of the system, defining the purpose and evaluation criteria, defining methods to achieve the goal, determining the structure of the system and its elements and achieving efficiency, quality and optimality of the system.

GIS-based statistics can complement consumer needs data and compensate for data gaps through complementarity. Analysis and visualization of the results in the studies are carried out using software for the study of global and local GIS.

The aim of the present study is to assess the possibilities of using geographic information systems to assess the potential of renewable energy sources in Ukraine, with substantiation and expansion of opportunities to use global and local GIS to assess the potential of renewable energy sources in Ukraine based on GIS analysis. The study will analyze key resource parameters in order to define areas for the development of renewable energy sources and assess their possible technical potential, as well as considering the effective integration of technologies for the use of renewable energy sources in the energy sector of Ukraine.

In the study, the estimation of possibilities of application of geoinformation systems for estimation of potential of renewable energy sources in Ukraine is executed with use of methodological bases and scientific results from a number of previous studies [16–22], and also using databases and interactive resource maps on global renewable energy sources from the internet [23–27].

The scientific novelty of our results lies in the application of global and local GIS for solving the problems of a comprehensive assessment of the potential and effectiveness of the use of regional non-traditional and renewable energy resources on the territory of Ukraine. This makes it possible to assess the possibilities of generating additional electric and thermal power for the needs of the regions of Ukraine using non-traditional and renewable energy sources. For the first time, a comprehensive methodology for the use of GIS is proposed for assessing the potential of non-traditional and renewable energy sources at the regional level in Ukraine, taking into account energy, environmental and socio-economic factors affecting the placement of non-traditional and renewable energy facilities [26,27]. The proposed principles provide the basis for rational search for and selection of territories that are most promising for the placement of facilities for non-traditional and renewable energy in Ukraine.

3. Results

In 2018, Ukraine became a member of the International Renewable Energy Agency (IRENA). According to [28,29], the activities of IRENA ensure coordination and intensification of the work of countries' agencies for renewable energy development through analysis, formulation of recommendations, and transfer of knowledge and technologies.

The authors used the IRENA database to obtain the results shown in Figures 1–3. Figure 1 shows the dynamics of growth of the installed capacity of technologies for the use of renewable energy sources in Europe. Figure 2 shows the dynamics and structure of growth of the installed capacity of technologies based on renewable energy sources in Europe in 2016–2020.

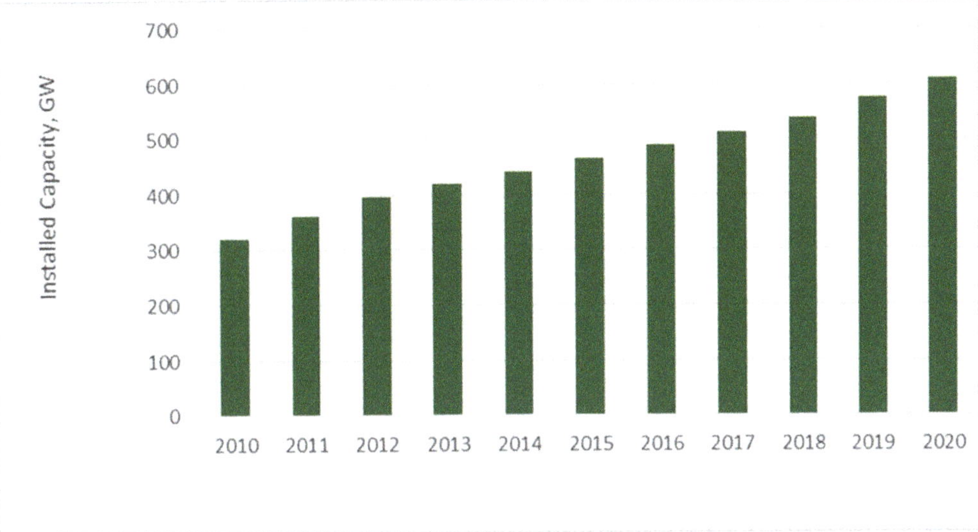

Figure 1. Dynamics of growth of the installed capacity of technologies for the use of renewable energy sources in Europe (IRENA).

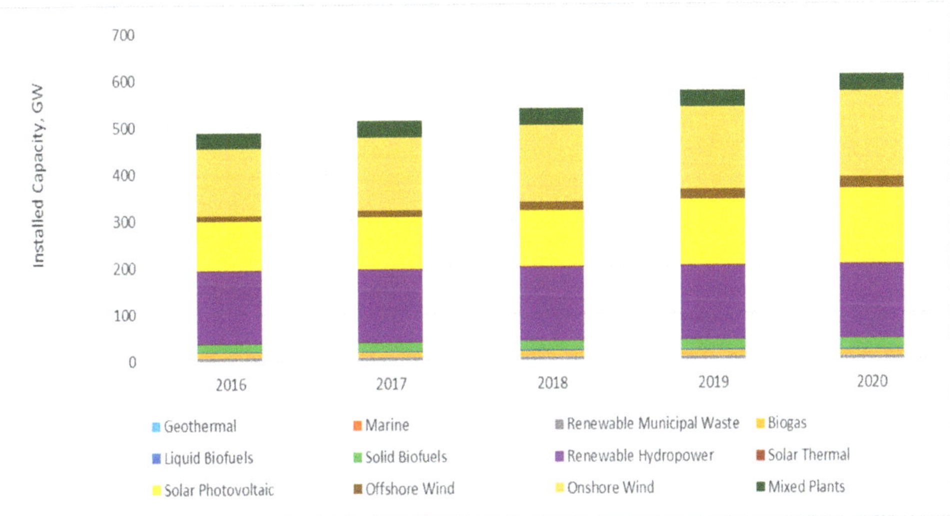

Figure 2. Dynamics and structure of growth of the installed capacity of technologies based on renewable energy sources in Europe in 2016–2020 (IRENA).

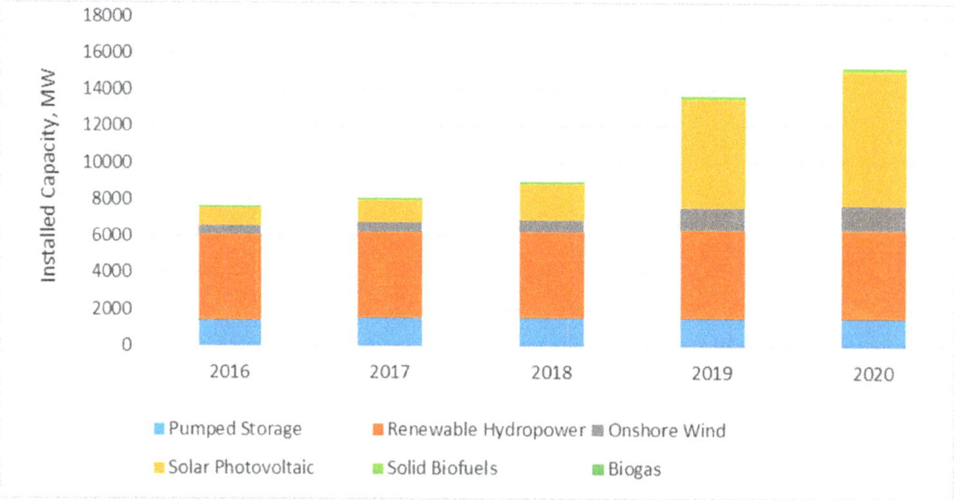

Figure 3. Dynamics and structure of growth of the installed capacity of technologies based on renewable energy sources in Ukraine in 2016–2020 (IRENA).

Figure 3 shows the dynamics and structure of growth of the installed capacity of technologies based on renewable energy sources in Ukraine in 2016–2020. IRENA member countries have the opportunity to use a number of software products of this organization, namely, statistics on the use of renewable energy (renewable capacity statistics) and the Global Atlas [30]. IRENA also offers a number of tools for the implementation of projects using renewable energy technologies: Project Navigator, Sustainable Energy Marketplace and IRENA/ADFD Project Facility.

The Global Atlas of the use of renewable energy sources is a web-based platform that provides users with access to renewable energy maps around the world.

More than 50 highly reputed international research institutes were involved in the creation of the Global Atlas, which provides more than 2000 maps with characteristics of renewable energy sources (solar, wind, bioenergy, geothermal and marine energy) on a single platform.

The Global Atlas zoning service provides GIS-based spatial analysis using the IRENA method for large-scale programs and applications, covering analysis of key parameters (quality of resources, distance to transmission networks, population density, terrain and protected areas) to determine suitable zones for the development of renewable energy sources and to give an approximate assessment of the technical potential.

IRENA software products are offered to support national and regional planning of renewable energy technologies in countries. The proposed technologies help to establish the viability of future projects for the use of renewable energy sources, and help project developers to characterize and analyze promising projects for the implementation of technologies using renewable energy.

Figures 4 and 5 show the selection of characteristics of renewable energy sources in the Global Atlas program from IRENA for any area with coordinates in Ukraine. This is based on reproduction of graphic material [30].

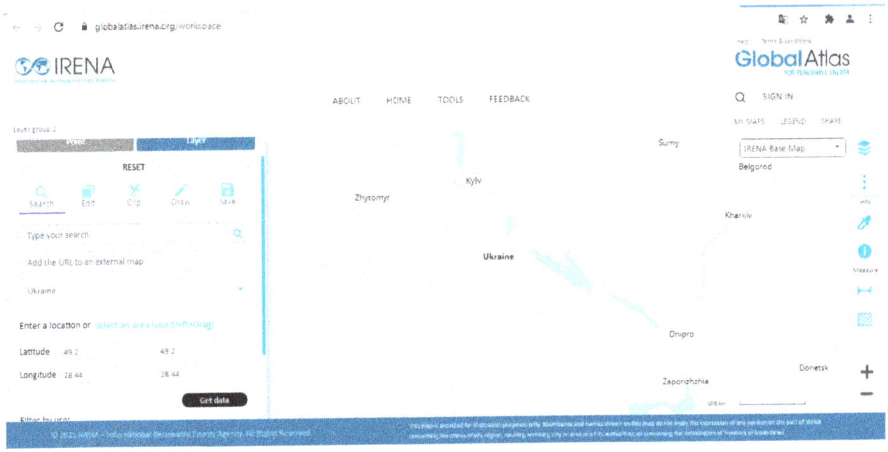

Figure 4. Selection of areas with coordinates in Ukraine to determine the characteristics of renewable energy sources in the Global Atlas program from IRENA (reproduction of graphic material from [30]).

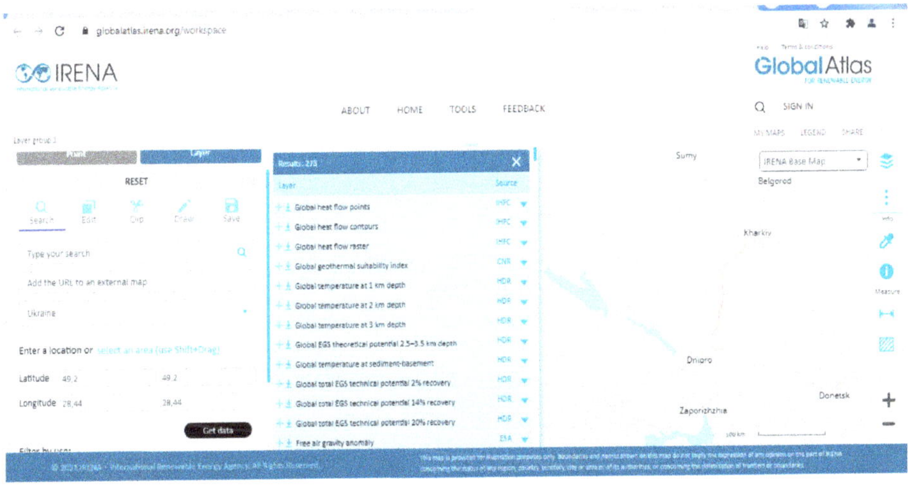

Figure 5. Selection of renewable energy characteristics in the Global Atlas program from IRENA for any area with coordinates in Ukraine (reproduction of graphic material from [30]).

4. Discussion

According to IRENA, based on the results of the analysis "REmap 2030" [31], Ukraine has significant potential for the development of solar and wind energy and the use of energy from biomass. According to the analytical review "REmap 2030", Ukraine can ensure a ten-fold increase in total end-use of renewable energy sources by 2030 (compared to 2009). According to forecasts, by 2030 the use of the potential of renewable energy sources in Ukraine will be distributed as follows: 73% of the potential will be provided in the heat industry, 20% in the electricity sector and 7% in the transport sector.

The use of Ukraine's additional potential in wind energy, biomass and solar photovoltaic will increase the share of energy produced from renewable sources in electricity production by 25% by 2030. In addition to IRENA web resources, it is advisable to use geographic information systems from Ukrainian web resources to assess the potential of renewable energy sources in Ukraine. In particular, «UA MAP» [32–37] is a modern information web resource containing information on renewable energy and energy efficiency in Ukraine. The interactive investment resource map of Ukraine posted on the «UA MAP» website (Figure 6) reflects general information on renewable energy projects in Ukraine and provides data on the resource potential of the area. Figure 6 reproduces graphic material in [34].

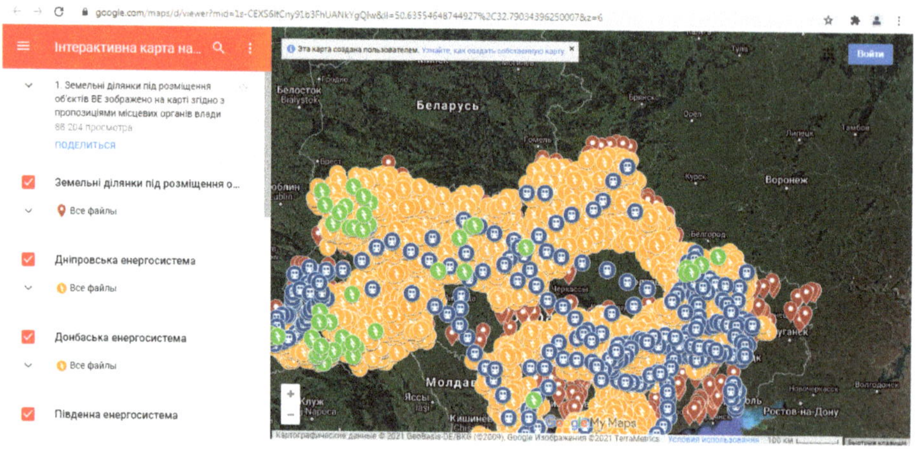

Figure 6. Interactive investment resource map of Ukraine "UA MAP" (reproduction of graphic material from [34]).

The «UA MAP» website contains an interactive online map of the energy sector of Ukraine (Figure 7), which shows the chains of energy transformations for the production of certain types of energy in Ukraine. Figure 7 is a reproduction of graphic material in [35].

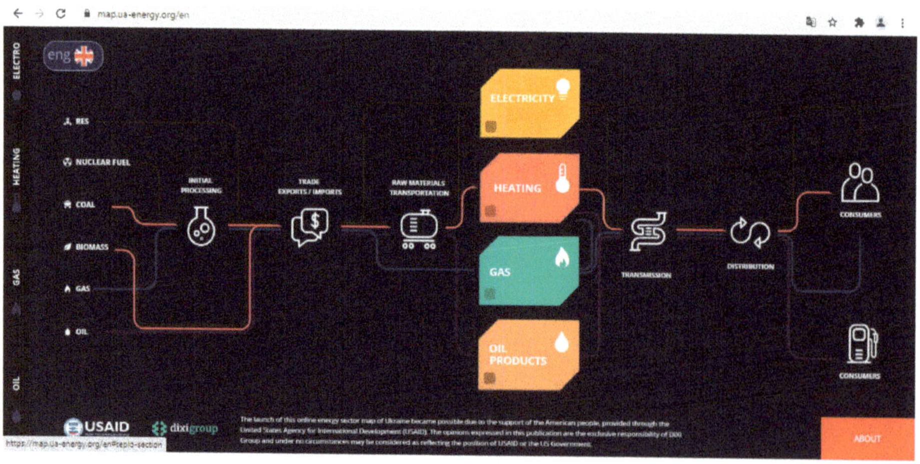

Figure 7. Interactive online map of the energy sector of Ukraine «UA MAP» (reproduction of graphic material from [35]).

Figure 8, for example, shows the chains of heat generation technologies in Ukraine based on traditional, non-traditional and renewable energy sources in the online map of the energy sector of Ukraine "UA MAP". Figure 8 is a reproduction of graphic material in [36]. In Figure 8, each tab provides outputs of statistical and balance data in Excel.

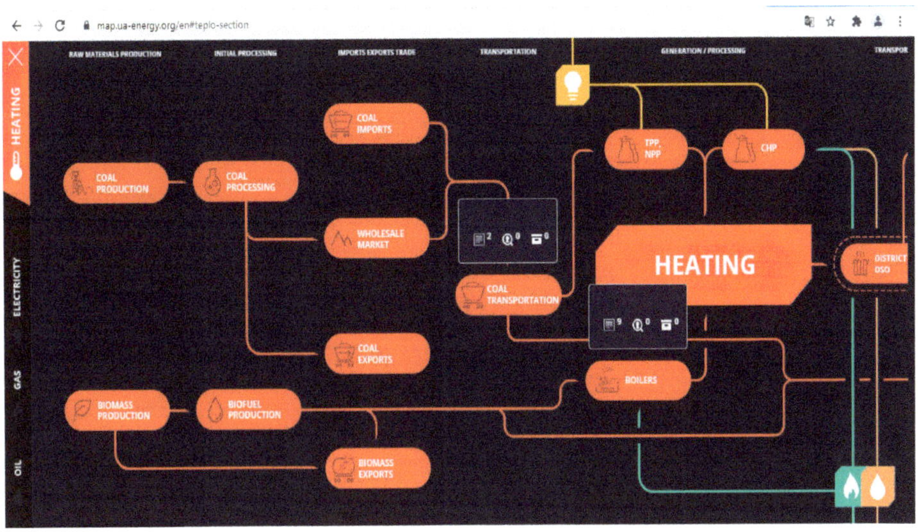

Figure 8. Chains of heat generation technologies in Ukraine based on traditional, non-traditional and renewable energy sources in the online map of the energy sector of Ukraine "UA MAP" (reproduction of graphic material from [36]).

The analyzed software products from the web resource "UA MAP" will permit assessment of the effective integration of certain technologies for the use of renewable energy sources in the energy sector of Ukraine.

The study shows the principles of using the generated vast data sets on the basis of the analyzed global and regional GIS systems, which are proposed to be used as a basis for research in the fields of non-traditional and renewable energy in Ukraine.

The study substantiates the scientific and methodological significance of GIS in studying the regional potential of non-traditional and renewable energy sources in Ukraine. This is of great practical use, since the analyzed databases and GIS should become an influential tool for analyzing the effectiveness of the practical use of non-traditional and renewable energy sources in the regions of Ukraine, as well as supporting the adoption of informed technical and managerial decisions on the use of potential regional non-traditional and renewable energy sources.

This paper study focuses on the need to create methods for using the resource potential of GIS in the field of non-traditional and renewable energy sources, taking into account the possible prerequisites and limitations for the development of non-traditional and renewable energy sources at the regional level. This approach makes it possible to carry out zoning of territories in order to select the most promising sites for projects for the development of non-traditional and renewable energy in the regions of Ukraine.

The study proposes the principles of using global and local GIS for solving the problems of a comprehensive assessment of the potential and efficiency of using regional non-traditional and renewable energy resources on the territory of Ukraine. This approach makes it possible to assess the possibilities of generating additional electrical and thermal power for the needs of the regions of Ukraine using non-traditional and renewable energy sources, to rationally assess the effectiveness of regional projects on the use of non-traditional and renewable energy sources in Ukraine.

The paper proposes a comprehensive methodology for using GIS to assess the potential of non-traditional and renewable energy sources at the regional level in Ukraine, taking into account energy, environmental and socio-economic factors affecting the placement of non-traditional and renewable energy facilities. The proposed principles provide a reasonable search and selection of territories that are most promising for the placement of facilities for non-traditional and renewable energy in Ukraine [38,39].

5. Conclusions

The scientific novelty of the results presented in this study is the substantiation and expansion of the possibilities of using global and local geographic information systems to assess the potential of renewable energy sources in Ukraine. GIS analysis aligned to key resource parameters permits identification of territories for development of renewable energy sources and assessment of their technical potential, as well as effective integration of technologies for the use of renewable energy sources in the energy sector of Ukraine.

The scientific novelty of the results obtained lies in the fact that the principles of applying global and local GIS make it possible to assess the possibilities of generating additional electric and thermal power for the needs of the regions of Ukraine using non-traditional and renewable energy sources, and to rationally assess the effectiveness of regional projects on the use of non-traditional and renewable energy sources in Ukraine. A comprehensive methodology for the use of GIS is proposed for assessing the potential of non-traditional and renewable energy sources at the regional level in Ukraine, taking into account energy, environmental and socio-economic factors affecting the placement of non-traditional and renewable energy facilities. The proposed principles provide for a rational search for and selection of territories that are most promising for the placement of facilities for non-traditional and renewable energy in Ukraine.

This paper analyzes the possibilities of using geographic information systems to assess the potential of renewable energy sources in Ukraine. The possibility of using the software products of the International Renewable Energy Agency, namely the Global Atlas, is analyzed.

This study confirms that the Global Atlas provides valuable spatial analysis based on GIS technologies using the IRENA method for large-scale programs and applications. This analysis of key parameters (resource quality, distance to transmission networks, population density, terrain and protected areas) permits determination of suitable areas for the development of renewable energy sources and gives an approximate assessment of the technical potential.

IRENA software products are offered to support national and regional planning of renewable energy technologies. The proposed technologies can help to establish the viability of future facilities for the use of renewable energy sources and help project developers to characterize and analyze promising facilities for the implementation of technologies for the use of renewable energy.

According to IRENA, based on the results of the analysis "REmap 2030", Ukraine has significant potential for the development of solar and wind energy, and the use of energy from biomass.

The application of geographic information systems from Ukrainian web resources to assess the potential of renewable energy sources in Ukraine is analyzed. The possibilities of the modern information web resource "UA MAP", which contains information on renewable energy and energy efficiency in Ukraine, are illustrated.

The study shows the principles of using the generated vast data sets on the basis of the analyzed global and regional GIS systems, which are proposed to be used as a basis for research in the fields of non-traditional and renewable energy in Ukraine.

The article substantiates the scientific and methodological significance of GIS in studying the regional potential of non-traditional and renewable energy sources in Ukraine, which is of great practical use. The analyzed databases and GIS should become an influential tool for analyzing the effectiveness of the practical use of non-traditional and renewable energy sources in the regions of Ukraine, as well as supporting the adoption of informed technical and managerial decisions on the use of the potential of regional non-traditional and renewable energy sources.

The study focuses on the need to create methods for using the resource potential of GIS in the field of non-traditional and renewable energy sources, taking into account the possible prerequisites and limitations for the development of non-traditional and renewable energy sources at the regional level. This approach makes it possible to carry out zoning of territories in order to select the most promising sites for projects for the development of non-traditional and renewable energy in the regions of Ukraine.

The study proposes the principles of using global and local GIS for solving the problems of a comprehensive assessment of the potential and efficiency of using regional non-traditional and renewable energy resources on the territory of Ukraine. For example, this approach makes it possible to better assess the possibilities of generating additional electrical and thermal power for the needs of the regions of Ukraine using non-traditional and renewable energy sources, through regional projects.

The study proposes a comprehensive methodology for using GIS to assess the potential of non-traditional and renewable energy sources at the regional level in Ukraine, taking into account energy, environmental and socio-economic factors affecting the placement of non-traditional and renewable energy facilities. The proposed principles provide for a rational search for, and selection of, territories that are most promising for the placement of facilities for non-traditional and renewable energy in Ukraine.

The analyzed software products will permit assessment of the effective integration of certain technologies for the use of renewable energy sources in the energy sector of Ukraine.

Author Contributions: Conceptualization, O.O. and V.K.; methodology, O.O.; software, O.O.; validation, O.O., P.O. and L.H.; formal analysis, D.M.; investigation, O.O., L.H. and O.P.; resources, V.K. and P.O.; data curation, O.O.; writing—original draft preparation, O.O. and V.K.; writing—review and editing, P.O. and D.M.; visualization, O.O.; supervision, L.H. and D.M.; project administration, P.O.; funding acquisition, L.H. All authors have read and agreed to the published version of the manuscript.

Funding: This research received no external funding.

Institutional Review Board Statement: Not applicable.

Informed Consent Statement: Not applicable.

Data Availability Statement: Not applicable.

Conflicts of Interest: The authors declare no conflict of interest.

References

1. Benalcazara, P.; Komorowska, A. Prospects of green hydrogen in Poland: A techno-economic analysis using a Monte Carlo approach. *Int. J. Hydrogen Energy* **2021**. [CrossRef]
2. Zatserkovnyi, V.; Oberemok, N.; Puzyk, A. Geoinformation modeling in the problems of renewable energy. *Bull. NTU "KhPI" Ser. New Solut. Mod. Technol. Kharkiv. NTU "KhPI"* **2018**, *9*, 118–127. [CrossRef]
3. Olczak, P.; Komorowska, A. An adjustable mounting rack or an additional PV panel? Cost and environmental analysis of a photovoltaic installation on a household: A case study in Poland. *Sustain. Energy Technol. Assess* **2021**, *47*, 101496. [CrossRef]
4. Cader, J.; Olczak, P.; Koneczna, R. Regional dependencies of interest in the "My Electricity" photovoltaic subsidy program in Poland. *Polityka Energetyczna Energy Policy J.* **2021**, *24*, 97–116. [CrossRef]
5. Arkhypova, L.M.; Mandryk, O.M.; Moskalchuk, N.M.; Prykhodko, M.M.; Radlovska, K.O. Renewable energy resources in the system of sustainable development of Carpathian region of Ukraine. *J. Phys. Conf. Ser.* **2021**, *1781*, 012010. [CrossRef]
6. Sabishchenko, O.; Rębilas, R.; Sczygiol, N.; Urbański, M. Ukraine energy sector management using hybrid renewable energy systems. *Energies* **2020**, *13*, 1776. [CrossRef]
7. Ostapenko, O.; Savina, N.; Marnatova, L.; Zienina-Bilichenko, A.; Selezneva, O. Perspectives of application of innovative resource-saving technologies in the concepts of green logistics and sustainable development. *Tour. Estud. Prat. (UERN)* **2020**, *2*, 1–12.
8. Ostapenko, O. Estimation of tendencies of transforming the energy sectors of World, European Union and Ukraine in the perspective to 2050 with using the renewable energy sources in the concept of Sustainable Development. In *Book Social Capital: Vectors of Development of Behavioral Economics: Collective Monograph*; ACCESS Press Publishing House: Veliko Tarnovo, Bulgaria, 2021; pp. 99–139.

9. Arkhypova, L.; Fomenko, N.; Kinash, I.; Golovnia, O. Territorial Recreational Systems and Sustainable Development. *Adv. Econ. Bus. Manag. Res.* **2019**, *99*, 189–194.
10. Mandryk, O.; Moskalchuk, N.; Arkhypova, L.; Prykhodko, M.; Pobigun, O. Prospects of environmentally safe use of renewable energy sources in the sustainable tourism development of the Carpathian region of Ukraine. *E3S Web Conf.* **2020**, *166*, 04005. [CrossRef]
11. Mandryk, O.; Arkhypova, L.; Pukish, A.; Zelmanovych, A.; Yakovlyuk, K. Theoretical and methodological foundations of sustainable development of Geosystems. *IOP Conf. Ser. Mater. Sci. Eng.* **2017**, *200*, 012018. [CrossRef]
12. Matuszewska, D.; Kuta, M.; Olczak, P. Techno-Economic Assessment of Mobilized Thermal Energy Storage System Using Geothermal Source in Polish Conditions. *Energies* **2020**, *13*, 3404. [CrossRef]
13. Rutitis, D.; Volkova, T. Model for Development of Innovative ICT Products at High-Growth Potential Startups. *Eurasian Stud. Bus. Econ.* **2021**, *19*, 229–241.
14. Nitsenko, V.S.; Mardani, A.; Streimikis, J.; Shkrabak, I.; Klopov, I.; Novomlynets, O.; Podolska, O. Criteria for Evaluation of Efficiency of Energy Transformation Based on Renewable Energy Sources. *Montenegrin J. Econ.* **2018**, *4*, 253–263.
15. Danyliuk, V.; Riepina, I.; Shafalyuk, O.; Kovylina, M.; Nitsenko, V. Functional and investment strategies of technical development of enterprises. *Nauk. Visnyk Natsionalnoho Hirnychoho Universytetu* **2020**, *3*, 115–121. [CrossRef]
16. Kinash, I.; Shtogryn, H.; Sakal, O.; Zapukhliak, I. The ecologization of housing and communal services of Ukraine in the context of sustainable development. *J. East. Eur. Cent. Asian Res.* **2019**, *6*, 113–130.
17. Kinash, I.; Arkhypova, L.; Polyanska, A.; Dzoba, O.; Andrusiv, U.; Iuras, I. Economic evaluation of tourism infrastructure development in Ukraine. *IOP Conf. Ser. Mater. Sci. Eng.* **2019**, *477*, 012020. [CrossRef]
18. Godfrey, B. *Renewable Energy: Power for a Sustainable Future*; Oxford University Press: Oxford, UK, 2012; p. 584.
19. Zelenko, Y.; Malovanyi, M.; Tarasova, L. Optimization of heat-and-power plants water purification. *Chem. Chem. Technol.* **2019**, *13*, 218–223. [CrossRef]
20. Mandryk, O.; Moskalchuk, N.; Arkhypova, L.; Pryhodko, M.; Pobigun, O. Research quantitative indicators of the potential of solar energy in the Carpathian region of Ukraine. *IOP Conf. Ser. Mater. Sci. Eng.* **2020**, *749*, 012033. [CrossRef]
21. Twidell, J.; Weir, T. *Renewable Energy Resources*, 3rd ed.; Routledge: London, UK, 2015.
22. Prykhodko, M.; Arkhypova, L.; Horal, L.; Kozhushko, S. Concept of ecosystem services and its implementation in Ukraine. *J. Geol. Geogr. Geoecol.* **2020**, *29*, 387–397. [CrossRef]
23. DataBank. WorldBank. Available online: https://databank.worldbank.org/reports.aspx?source=2&country=UKR# (accessed on 2 December 2021).
24. Esmap. Available online: https://www.esmap.org/ (accessed on 2 December 2021).
25. Eurostat. Available online: https://ec.europa.eu/eurostat/data/database (accessed on 2 December 2021).
26. Eurostat. Energy Dependency. Available online: https://ec.europa.eu/eurostat/statistics-explained/index.php?title=Energy_statistics_-_an_overview#Energy_dependency (accessed on 2 December 2021).
27. Tracking SDG7. Available online: URL:https://trackingsdg7.esmap.org/country/ukraine (accessed on 2 December 2021).
28. International Renewable Energy Agency (IRENA). Available online: https://www.irena.org/ (accessed on 2 December 2021).
29. Statute of International Renewable Energy Agency (IRENA). Available online: https://irena.org/-/media/Files/IRENA/Agency/About-IRENA/Statute/IRENA_FC_Statute_signed_in_Bonn_26_01_2009_incl_declaration_on_further_authentic_versions.pdf?la=en&hash=635C494208DD405EA8CD2BDB04414FECD40F55F1 (accessed on 2 December 2021).
30. Global Atlas of International Renewable Energy Agency (IRENA). Available online: https://www.irena.org/globalatlas (accessed on 2 December 2021).
31. Renewable Energy Roadmap of International Renewable Energy Agency (IRENA). Available online: https://saee.gov.ua/sites/default/files/ENG%20IRENA_REmap_Ukraine_paper_2015%201304.pdf (accessed on 2 December 2021).
32. Ukrainian Cartographic Network UA MAP. Available online: https://uamap.org.ua/ (accessed on 2 December 2021).
33. Ukrainian Cartographic Network UA MAP. Resource Maps. Available online: https://uamap.org.ua/resursnee-karti (accessed on 2 December 2021).
34. Ukrainian Cartographic Network UA MAP. Interactive Investment Resource Map. Available online: https://www.google.com/maps/d/viewer?mid=1s-CEXS6ltCny91b3FhUANkYgQIw&ll=50.63554648744927%2C32.79034396250006&z=6 (accessed on 2 December 2021).
35. Ukrainian Cartographic Network UA MAP. Interactive Online Map of the Energy Sector. Available online: https://map.ua-energy.org/en (accessed on 2 December 2021).
36. Ukrainian Cartographic Network UA MAP. Chains of Heat Generation in the Online Map of the Energy Sector. Available online: https://map.ua-energy.org/en#teplo-section (accessed on 2 December 2021).
37. Ukrainian Cartographic Network UA MAP. Chains of Heat Generation from CHP and NPP in the Online Map of the Energy Sector. Available online: https://map.ua-energy.org/en/categories/teplo/tes-aes (accessed on 2 December 2021).
38. Sribna, Y.; Koval, V.; Olczak, P.; Bizonych, D.; Matuszewska, D.; Shtyrov, O. Forecasting solar generation in energy systems to accelerate the implementation of sustainable economic development. *Polityka Energetyczna—Energy Policy J.* **2021**, *24*, 5–28. [CrossRef]
39. Koval, V.; Sribna, Y.; Kaczmarzewski, S.; Shapovalova, A.; Stupnytskyi, V. Regulatory policy of renewable energy sources in the European national economies. *Polityka Energetyczna—Energy Policy J.* **2021**, *24*, 61–78. [CrossRef]

MDPI
St. Alban-Anlage 66
4052 Basel
Switzerland
www.mdpi.com

Applied Sciences Editorial Office
E-mail: applsci@mdpi.com
www.mdpi.com/journal/applsci